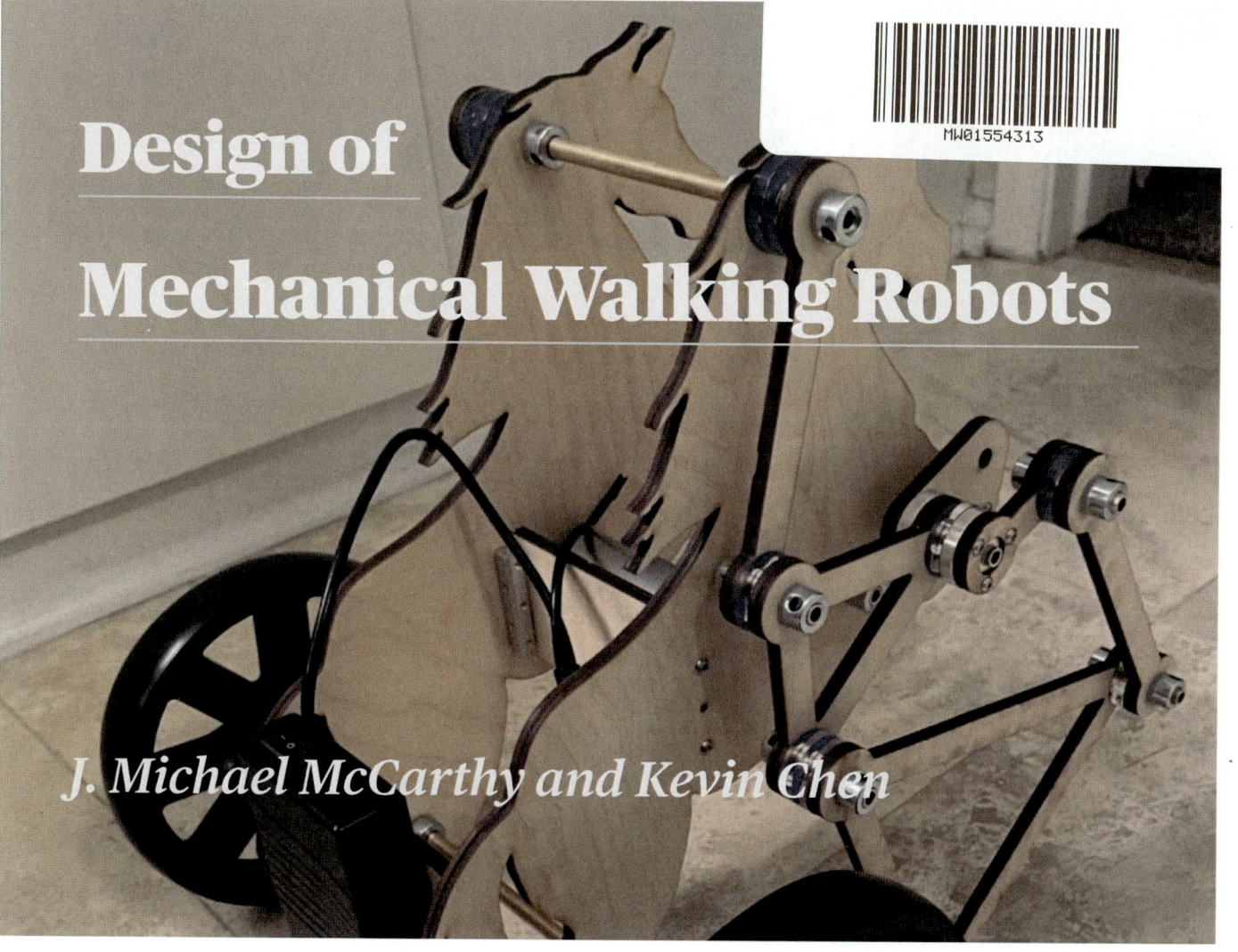

Design of Mechanical Walking Robots

J. Michael McCarthy and Kevin Chen

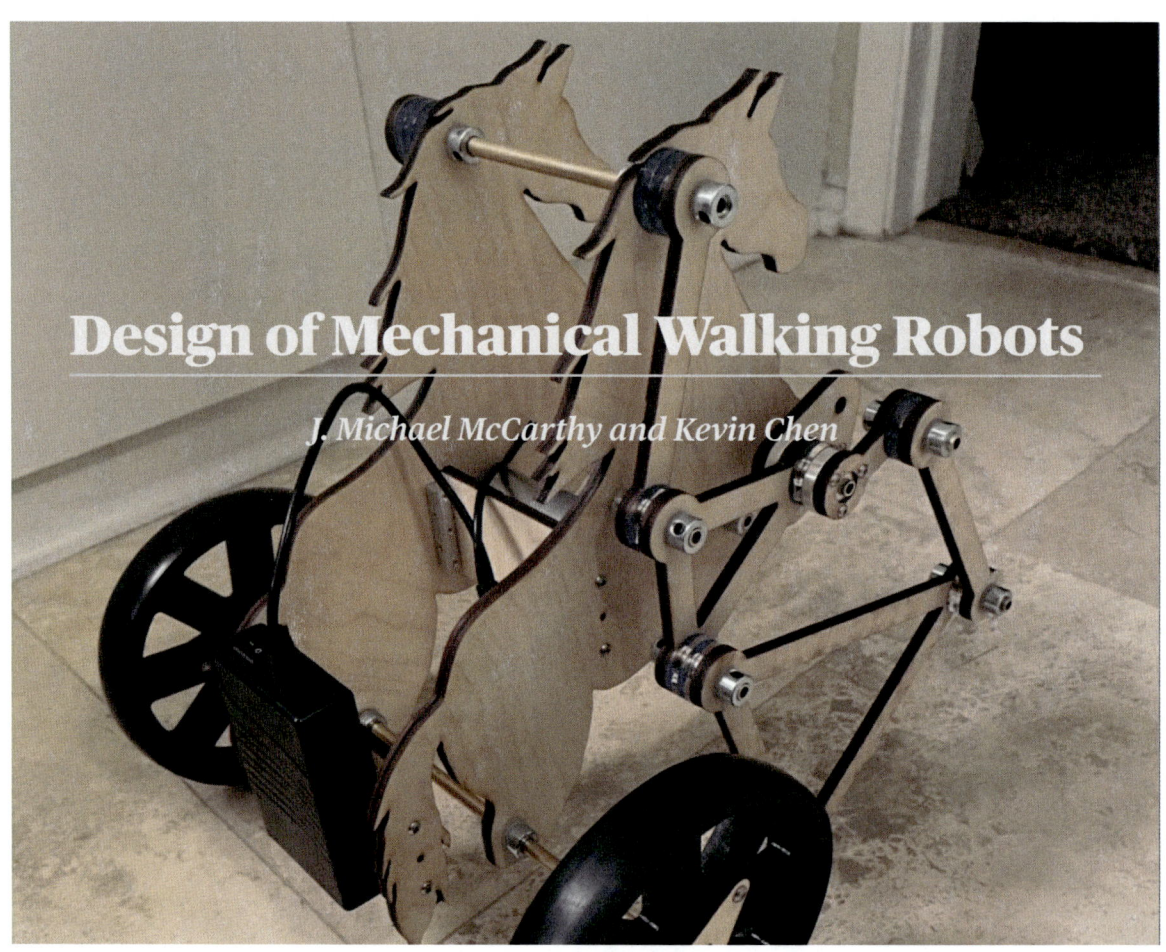

Design of Mechanical Walking Robots

J. Michael McCarthy and Kevin Chen

Design of Mechanical Walking Robots

© MDA Press, Irvine, CA, 2021. All rights reserved.
ISBN: 978-0-9785180-6-6

Dedication

This book is dedicated to the creativity, commitment, and enthusiasm for the design of mechanical walking robots shown by our student teams.

Preface

In this book, we present the detailed design of mechanical walking robots that are driven by a single motor. These walkers rely on specially designed leg mechanisms coordinated by gear trains in order to walk, rather than using multiple servomotors on each leg. Several of the walkers include simple control electronics and software for drive motor speed and steering angle. The result is a simplified walking robot that provides a platform for other mechanical and electronic functions.

Two, four and six legged walkers are presented each of which implements different types of leg mechanisms, power trains and control electronics. In each case, we provide drawings for a laser cut wood or acrylic chassis, 3D printed parts and a complete parts list. Our goal is to provide enthusiasts of all backgrounds what they need to build a walking robot at home, to explore new design ideas, and, perhaps, to enjoy the operation of one of these robots as it moves across the ground.

The walkers we describe are the result of design studies by teams of undergraduate and graduate students in the Department of Mechanical and Aerospace Engineering at the University of California, Irvine.

J. Michael McCarthy and Kevin Chen, Irvine 2021

Table of Contents

Preface	vi
Chapter 1. Overview	1
— Design Studies, Digital and Physical Prototypes, Foot Trajectories	
Chapter 2. Two Legged Delorean	5
— Chassis, Leg Mechanism, Power Train, Electronics, Software, Parts, Feet	
Chapter 3. Two Legged Ostrich	20
— Chassis, Leg Mechanism, Power Train, Electronics, Software, Parts, Feet	
Chapter 4. Four Legged Nightmare	38
— Chassis, Leg Mechanism, Power Train, Parts	
Chapter 5. Four Legged Chameleon	50
— Chassis, Leg Mechanism, Power Train, Electronics, Software, Parts, Feet	
Chapter 6. Six Legged Anteater	67
— Chassis, Leg Mechanism, Power Train, Electronics, Software, Parts, Walking	
Chapter 7. Conclusion	82
— Summary, Future Work, Flyers and Swimmers	
References	89
Index	90
About the Authors	91
Descriptions and Credits	92

CHAPTER 1

Overview

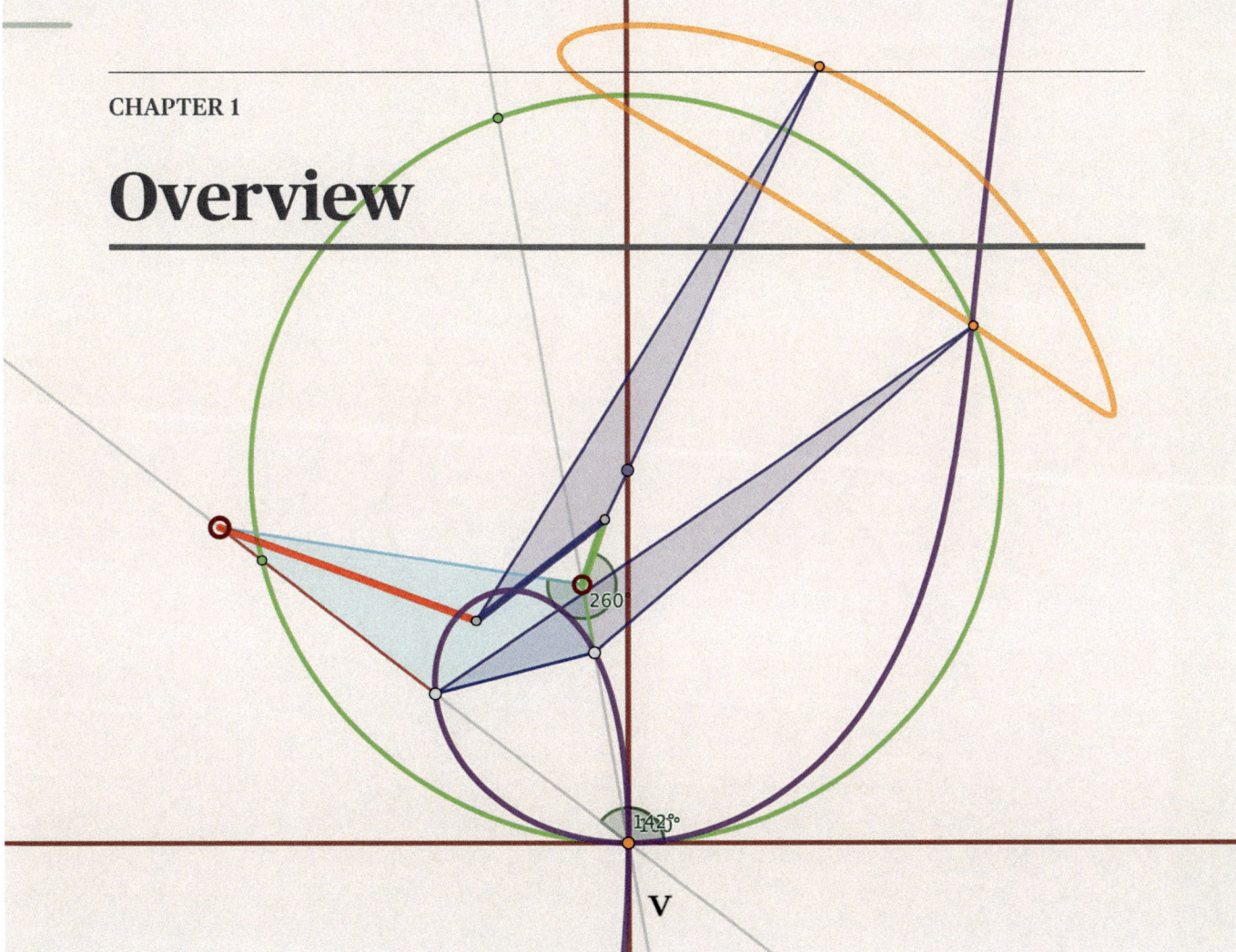

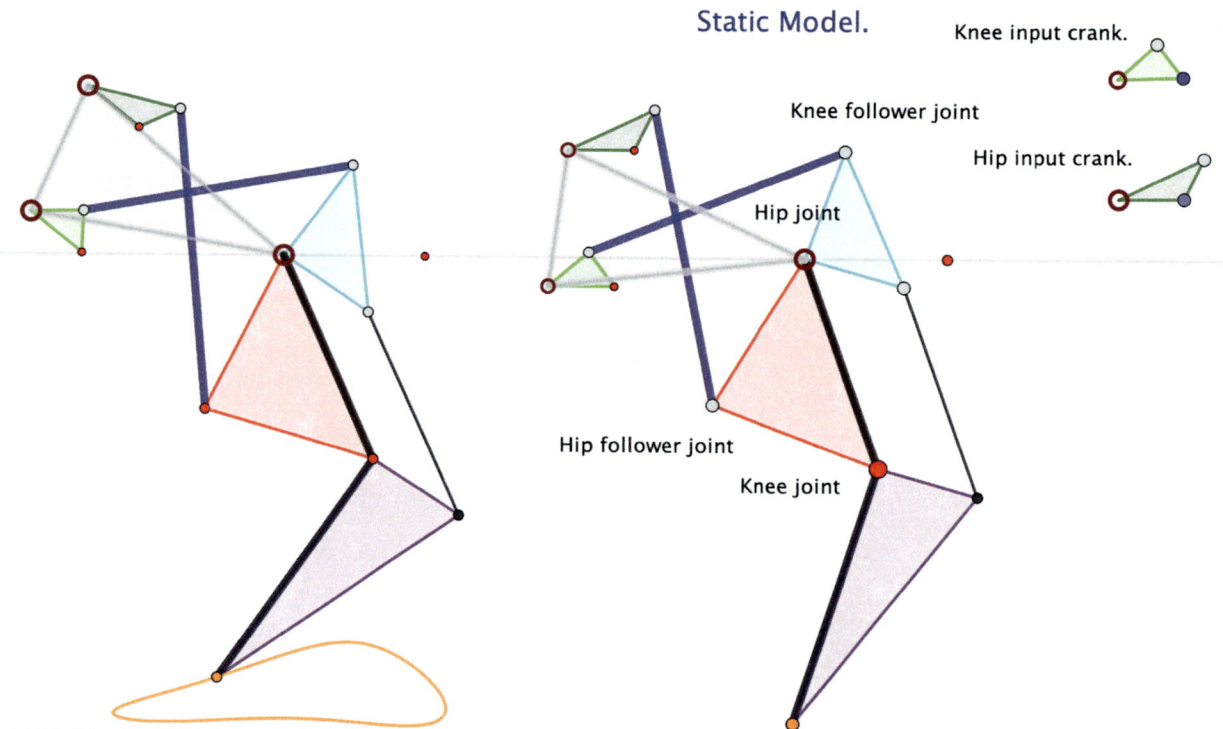

Jansen style leg mechanism. This is a generalization of Theo Jansen's leg design developed for his Strandbeest. The adjustments to the dimensions of the static model change the movement of the leg and the foot trajectory in the kinematic model.

Design Studies
Digital and Physical Prototypes

A walking robot propels itself forward with legs. The leg mechanisms of our walking robots generate cyclic foot trajectories and are interconnected with gear trains so they are driven by a single motor. The legs contact the ground either in phase or out of phase to provide a walking movement.

This means our walking machines are simpler than robot walkers that have servomotors on each leg joint, but they are more complex than wheeled and tracked vehicles that also have one drive motor. Several of our walkers include a separate steering motor.

The foot trajectory is preplanned and built into the leg mechanism, and therefore does not need computer control. It is possible to make the foot trajectory adjustable but that is a topic for future work.

The computer control of these walkers is simple and consists of speed control for the drive motor and position control of a steering servomotor.

The choice of leg mechanisms for these walkers is not so simple. The basic shape of the foot trajectory can be found in Shigley (1960) (see References), which consists of a flat-sided oval. Different versions of this shape can be traced by the coupler curve of a four-bar linkage, and then positioned relative to the body of the walker by means of a pantograph.

Geometric constructions for four-bar linkages with flat-sided coupler curves are presented in McCarthy (2019); also see Dijksman (1976) and

Foot Trajectory

Shieh (1996). These four-bar linkages can be modified using a pantograph, a skew pantograph or a rectilinear link to achieve leg mechanisms that provide useful features for the walking movement.

Another way to generate a flat-sided oval is to use four-bar function generators to drive the hip and knee joints of a simple leg. This is described by Chen and McCarthy (2021), who obtained a generalized version of the leg mechanism used by Theo Jansen in his Strandbeest, Jansen (2016). The generalized Jansen leg mechanism introduces a parallelogram along the upper leg, where Jansen has dimensions that differ from a parallelogram.

Once a leg mechanism is selected the joint design, chassis, drive train and electronics are needed to complete the walker. The joints of the leg mechanism must allow low friction movement and resist out-of-plane leg bending. Significant lateral forces arise as the vehicle shifts its load from one leg to the other.

The design of the chassis is an opportunity to demonstrate creativity, but it must be stiff enough to support torsion forces during walking.

Gearing is a convenient way to connect the drive motor to the input cranks of the legs. Our gears are manufactured by 3D printing or laser cutting using standard dimensions for involute gear teeth.

Our electronic control systems are based on the Arduino and provide speed control of the drive motor and position control of the steering servomotor. We have used both radio-frequency transmitters, as well as infrared transmitters to send commands to our walkers.

In what follows, digital and physical prototypes, laser cut drawings and hardware parts lists are illustrated for a range of different walkers. The goal is to assist the enthusiast in exploring and advancing the design and fabrication of mechanical walking robots.

CHAPTER 2

The Two Legged Delorean

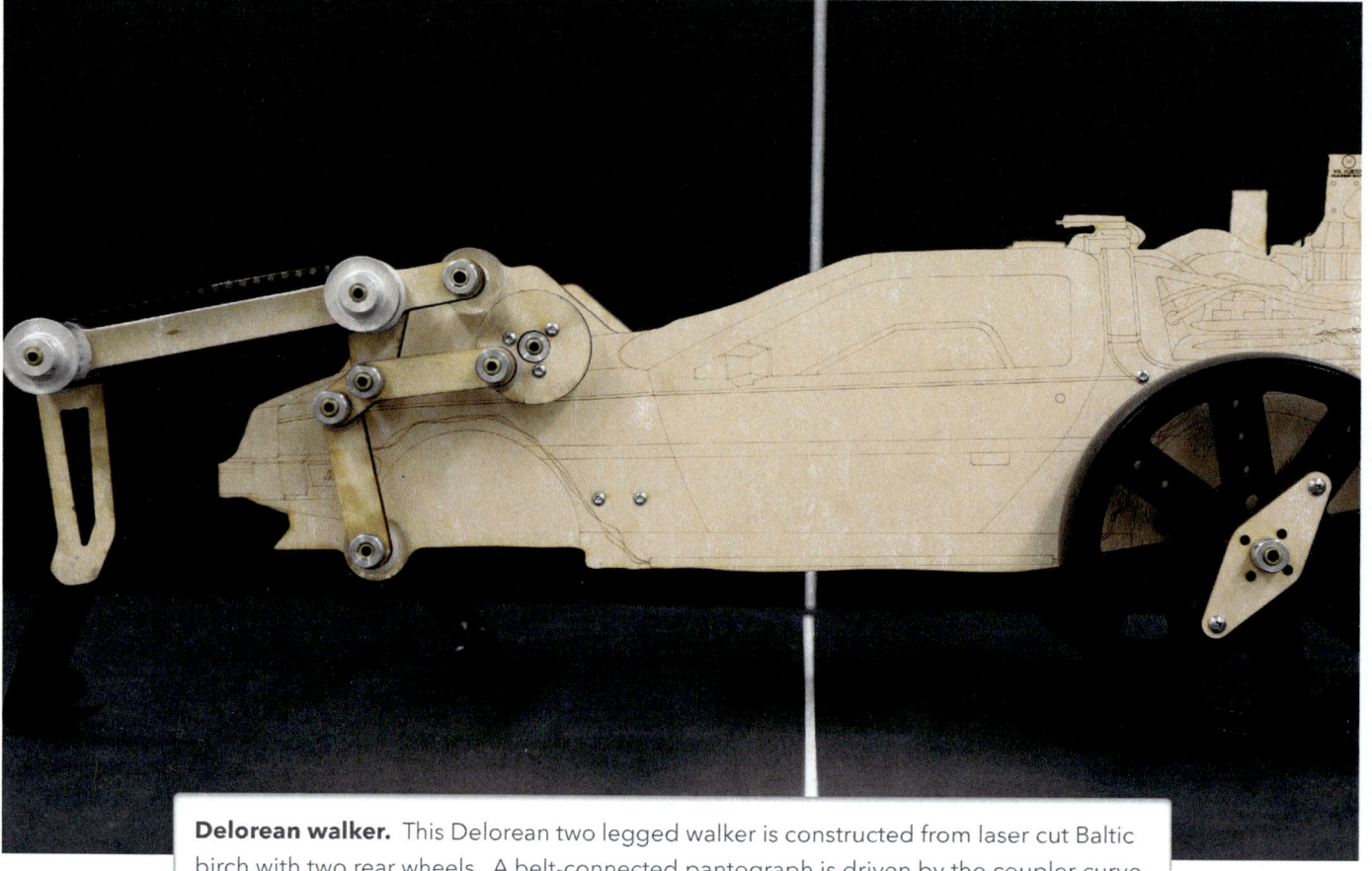

Delorean walker. This Delorean two legged walker is constructed from laser cut Baltic birch with two rear wheels. A belt-connected pantograph is driven by the coupler curve of a four-bar linkage to generate the foot trajectories for the two front legs.

Section 1
Chassis: Baltic Birch Delorean

The design of mechanical walking robot begins with the choice of the number of legs and the chassis that supports the legs. The chassis consists of two primary side members connected by a cross members. The construction of these members by laser cutting a sheet of material makes it easy to give these walkers personality.

Our Delorean walker was designed to explore the implementation of a four-bar linkage driven pantograph leg mechanism with speed control. Therefore we chose a chassis design that supports two front legs and two rear wheels without steering.

The walker chassis is constructed from 6mm Baltic birch with four cross members to form a stiff structure, in order to support the the alternating loads that arise as the legs contact the ground.

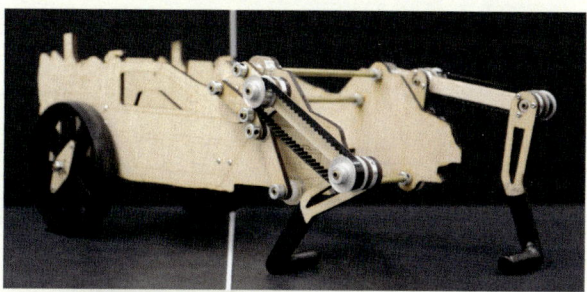

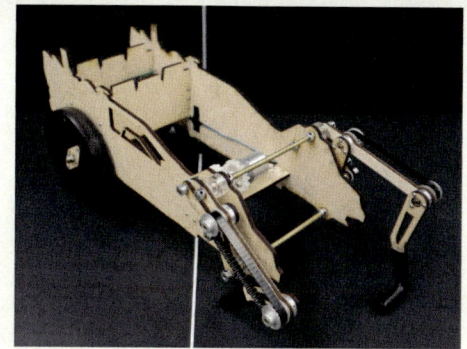

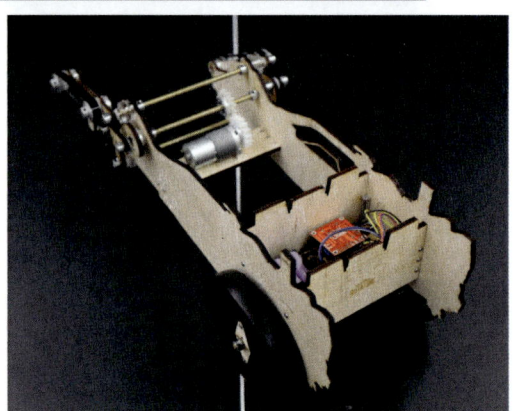

Section 2
Leg Mechanism: Four-bar Driven Pantograph

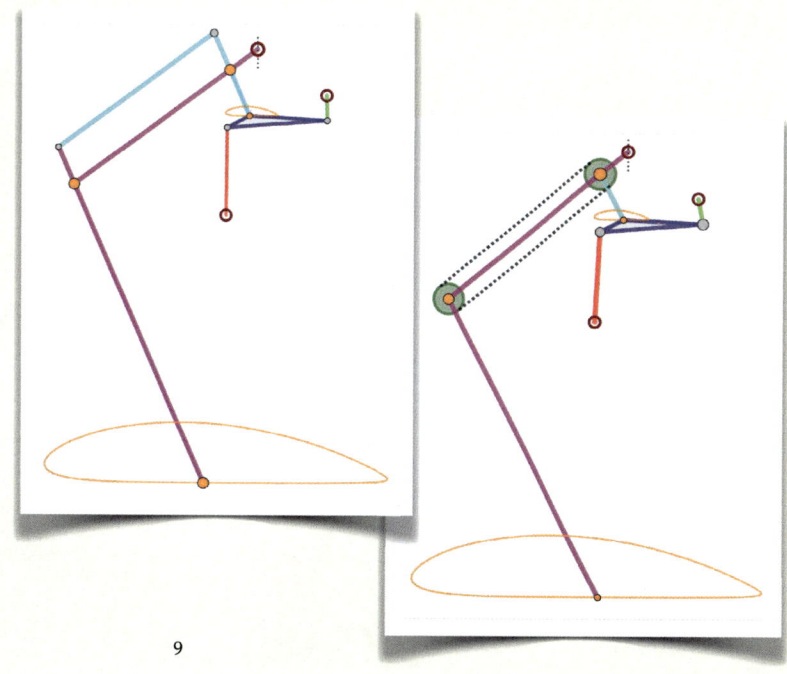

The kinematic structure of the leg mechanism for the Delorean was designed using Geogebra Classic, see geogebra.org.

This was a three step process:
1. Design a four-bar linkage that has a desired flat-sided coupler curve;
2. Introduce a pantograph linkage to expand the size of the coupler curve, and
3. Introduce a belt drive to replace the parallelogram link in order to improve performance.

Digital Prototype

We have found time and again that it is worthwhile to make a complete digital prototype of the walking machine using actual part models of available from suppliers.

The design of the leg joints are an ongoing concern. We have explored with and without bushings, nylon and brass bushings, shaft collars and screws, brass tubes or shoulder bolts. In every case, the digital model has helped avoid errors.

An accurate digital model also allows simulation of the walking movement, which helps understanding of the final performance. It also shows how the parts come together during assembly.

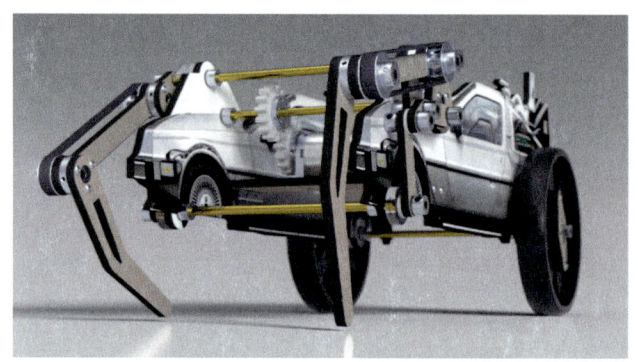

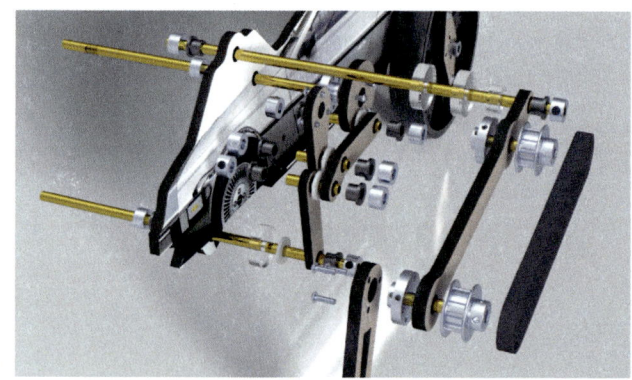

Physical Prototype

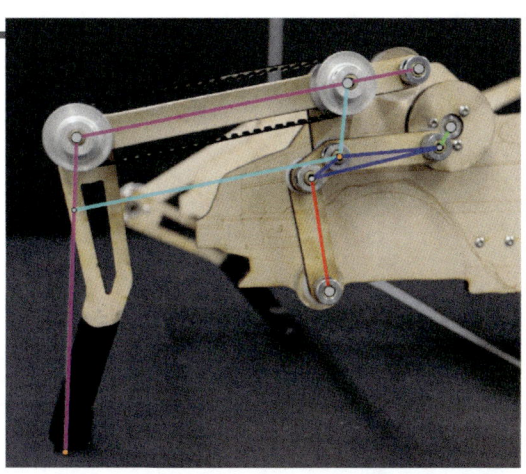

The chassis and leg components are cut from single sheets of 6mm Baltic birch using a laser cutter. The hardware consists of parts that are readily available on-line.

We have found it useful to use hubs, shaft collars and nut strips in order to provide structural stiffness and reliability, though they are more expensive.

The Delorean walker is 57 cm (22.5 in) long and 27 cm (10.5 in) wide and weighs 1.90 kg.

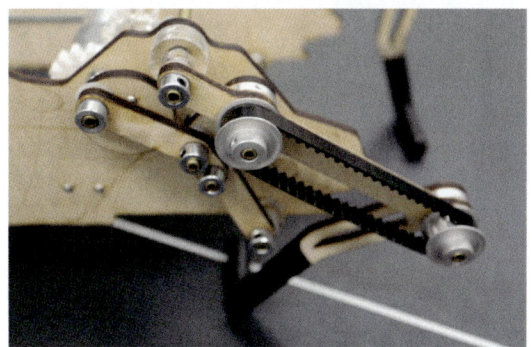

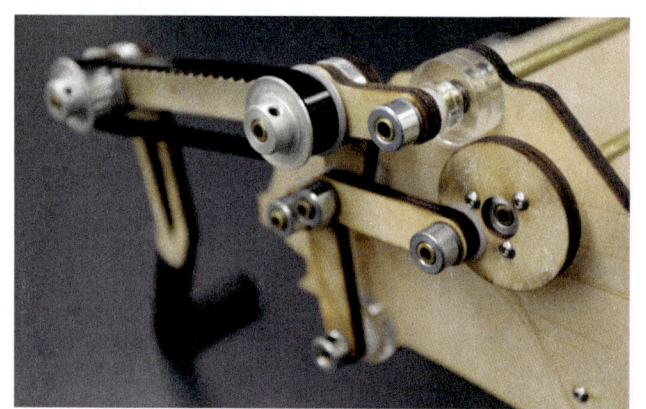

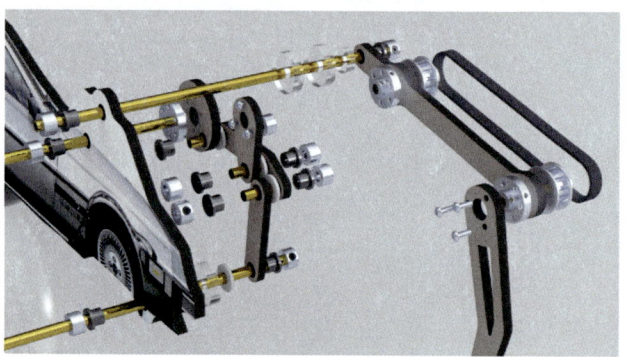

Section 3
Power Train: One Motor with Acrylic Drive Gears

The power train for the Delorean walker consists of a motor connected through a gear train to a drive axle formed by a brass rod. Hubs connect the drive axle to the input cranks for the two leg mechanisms.

A 12V battery pack drives the DC motor. The gears are laser cut from 1/4 inch acrylic. Hubs are used to connect the gears to the motor shaft and to the brass drive axle.

The drive axle is mounted to the chassis side panel by brass bushings to reduce friction. Shaft collars mounted on the drive axle are used to support side loading. Hubs mounted at the end of the drive axle connect to the input cranks of the leg mechanisms.

The two input cranks are positioned at an angle of 180 degrees in order to ensure alternating contact of the legs with the ground.

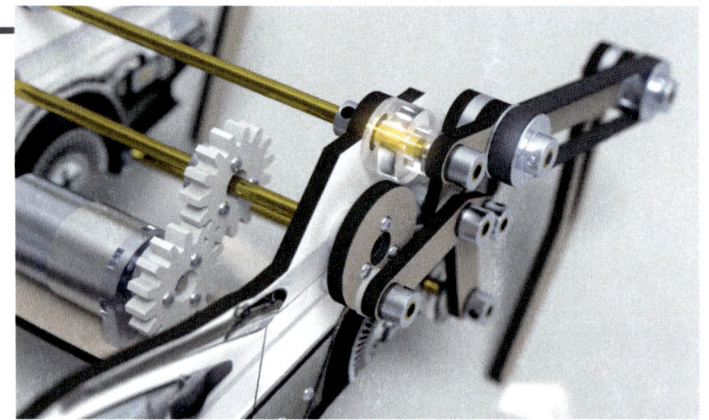
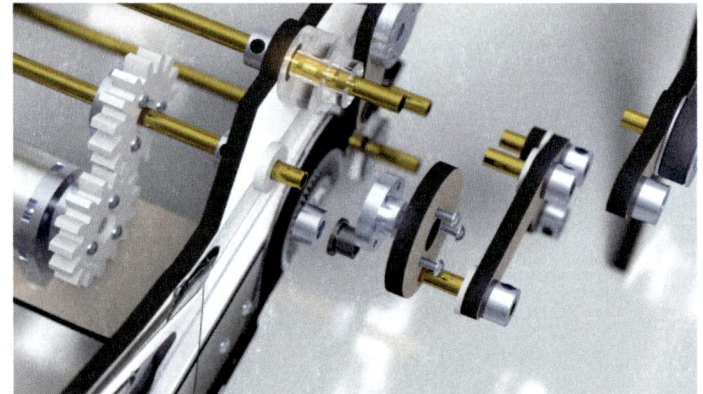

Section 4
Electronics: Speed Control with an RF Transmitter

The current to the DC motor is controlled by a signal from a transmitter used for remote controlled cars. The signal from the RF transmitter is received by a circuit that includes an Arduino Nano. A program in the Nano interprets the signal and controls the current from the batter to the motor through an H-bridge. The schematic for this circuit is shown in the figure.

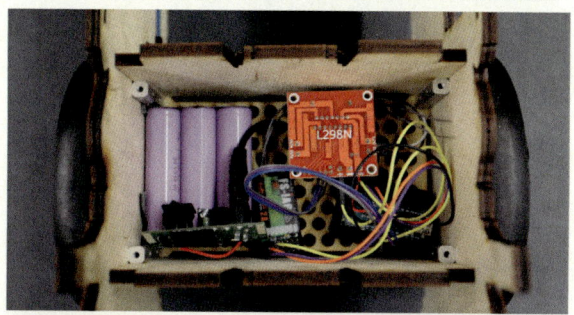

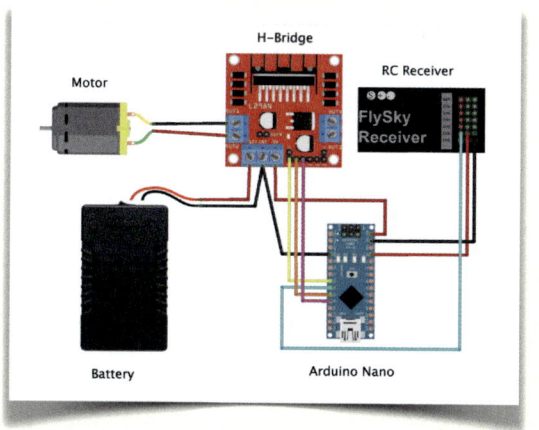

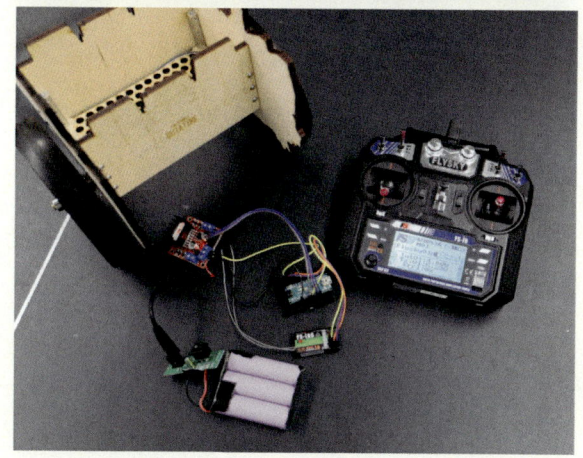

Software

The Arduino program consists of a loop that reads the RF transmitter signal and computes the output to the pulse width modulator (PWM) based on the joystick position, which then powers the DC Motor through the H-bridge.

[Flowchart: Define variables and parameters and run set-up. → Drive loop ↔ RF Receiver, Drive loop → DC Motor Drive]

```
L298N_FS-i6_FS-iA6 §
//Variables
int channel_2_value;

//Parameters
const int deadzone = 20;

//Receiver Pin
const int CH2_PIN = 6;

//Motor Driver Pins
const int IN1 = 2;
const int IN2 = 3;
const int ENA = 5;

void setup() {
    pinMode(CH2_PIN, INPUT);
    pinMode(IN1, OUTPUT);
    pinMode(IN2, OUTPUT);
    pinMode(ENA, OUTPUT);
    Serial.begin(9600);
}
```

```
void loop() {

    //read pulse-width from receiver
    channel_2_value = pulseIn(CH2_PIN, HIGH);

    //Convert to PWM
    channel_2_value = receivertoPWM(channel_2_value);

    //Drive the motor
    drive();

    delay(5);

//    //For debugging
//    Serial.println(channel_2_value);
}
```

```
void drive() {

    //Stops motor if no input from controller
    if (channel_2_value == 0) {
        digitalWrite(IN1, LOW);
        digitalWrite(IN2, LOW);
    }
    //Sets forward direction for H-Bridge
    //if throttle is pushed up on the controller
    else if (channel_2_value > 0) {
        digitalWrite(IN1, LOW);
        digitalWrite(IN2, HIGH);
    }
    //Sets backward direction for H-Bridge
    //if throttle is pulled down on the controller
    else if (channel_2_value < 0) {
        digitalWrite(IN1, HIGH);
        digitalWrite(IN2, LOW);
    }
    //Cuts power from H-Bridge for anything else
    else {
        digitalWrite(IN1, LOW);
        digitalWrite(IN2, LOW);
    }

    //Set motor speed
    analogWrite(ENA, abs(channel_2_value));
}
```

```
//Function: convert and constrain pulse-width
//values from receiver to PWM, return integer to loop
int receivertoPWM(int pulsefromRC) {

    //If we are getting a signal from
    //receiver and controller is turned on
    //lowest number when throttle is pulled back is 993
    //when controller is off it is 0
    if (pulsefromRC > 950) {
        //Map receiving value from 1010 (low) to 1965 (high)
        //to -255 (low) to 255 (high)
        //1010 to 1965 is controller FS-i6 and receiver FS-iA6 specific
        pulsefromRC = map(channel_2_value, 1010, 1965, -255, 255);

        //Remove slop when controller is at max and min
        pulsefromRC = constrain(pulsefromRC, -255, 255);
    }
    //If controller is turned off
    else {
        pulsefromRC = 0;
    }

    //Values in deadzone will stop the motor
    if (abs(pulsefromRC) <= deadzone) {
        pulsefromRC = 0;
    }
}
```

Section 5
Laser Cut Parts

The side and cross members of the chassis and the links of the leg mechanisms are cut from 12 x 20 inch sheets of 6mm Baltic birch.

The red lines indicate where the laser cuts completely through the wood to make individual parts. The dark lines denote surface etching to create an image.

Circular spacers 3/4 in OD and 1/4 ID are cut from 1/4 inch acrylic, to provide spacing and lateral support between the links of the leg mechanism.

Acrylic spacers move smoothly over each other, but the wood of the links should be sanded smooth where it contacts a spacer or a shaft collar to reduce friction.

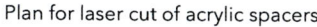
Plan for laser cut of acrylic spacers

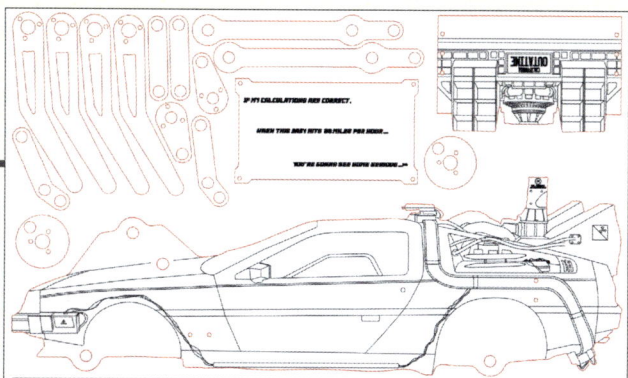
Plan for laser cut of leg links and a side and back of the walker.

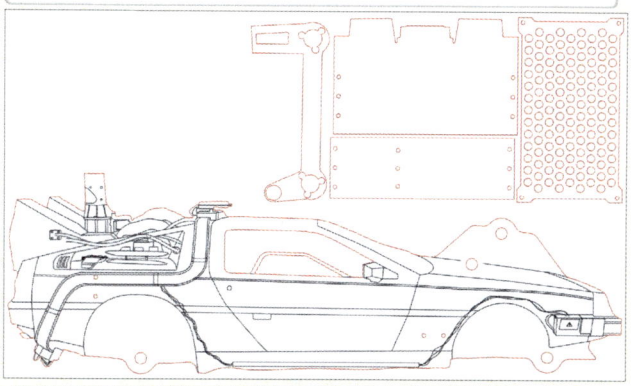
Plan for laser cut of the other side and bottom of the walker.

Parts List

Parts for the Delorean walker were selected for their availability orders with only days required for delivery. And, the goal was that assembly would only require hand tools.

In particular, a hack saw and file was used to size the brass tube and screw drivers were used to assemble the shaft collars, nut strips and hubs. Wiring the motor requires soldering two wires. The electronics was assembled without soldering.

The result is the components are expensive, but assembly is easy.

Description	Amount		Cost
Legs and Chassis Parts			
Nylon washers, 1/4 in screw size, 0.281 ID, 0.734	1	pkg of 50	$14.33 $14.33
Shaft Collar, 1/4 in dia, zinc plated 1215 steel.	24	each	$1.20 $28.80
Ultra-low friction oil-embedded sleeve bearing, flanged, 1/4 in did, 3/8 in housing, 3/16 in long	24	each	$0.92 $22.08
Brass Tubing, 1/4 in OD, 0.32 in wall thickness	6	feet	$7.05 $42.30
M3 10 mm long screw	1	pkg of 100	$6.93 $6.93
Timing Belt Drive Parts			
Corrosion-Resistant Timing Belt Pulley, XL Series, for 3/8" Width, with Hub, 2 Flanges, 1.094" OD	4	each	$9.62 $38.48
Dust-Free Timing Belt, XL Series, 3/8" Width, Trade No. 120xL037	2	each	$4.58 $9.16
Motor Assembly Parts			
Mounting Hub, 1/4 inch shaft, M3 holes	5	pkg of 2	$7.95 $39.75
Mounting Hub, 6mm shaft, M3 holes	1	pkg of 2	$7.95 $7.95
37D Motor (Polulu) 100:1	1	each	$24.95 $24.95
37D Motor Bracket (Pololu)	1	each	$7.95 $7.95
Hex Drive Screw, M3 x 0.5mm 10mm long	1	pkg of 100	$6.93 $6.93
Nut Strip 40mm for M3 screws	6	each	$2.25 $13.50
Wheel Assembly Parts			
Heavy Duty Wheel, 6 inch, (Servocity)	2	each	$14.33 $28.66
Wheel adapter screw, Extra wide phillips head, M3 x 0.5 mm, 18 mm long	1	pkg of 100	$4.74 $4.74
Wheel adapter nut, M3 x 0.5 mm thread	1	pkg of 100	$4.74 $4.74
Electronics Assembly Parts			
12V DC Lithium ion battery, 3000mAh	1	each	$24.79 $24.79
Arduino Nano	1	each	$19.80 $19.80
L298 H-bridge motor driver	1	pkg of 5	$10.89 $10.89
Flysky FS-i6X 2.4 GHz RC Transmitter w. Receiver	1	each	$53.99 $53.99
Mini-bread board	1	pkg of 6	$5.99 $5.99
Total			$416.71

Section 6
Feet

Contact between each leg and the ground provides alternating support for the center of gravity and propelling forces for forward and rearward movement. For the Delorean walker, we added a cylinder that extends inward from the end of the leg to form a foot.

The foot and portions of the leg were coated in Plasti Dip rubber coating to provide frictional contact for movement on linoleum and hardwood floors. We have obtained effective walking movement without specialized feet, however, feet can improve performance. and deserve attention.

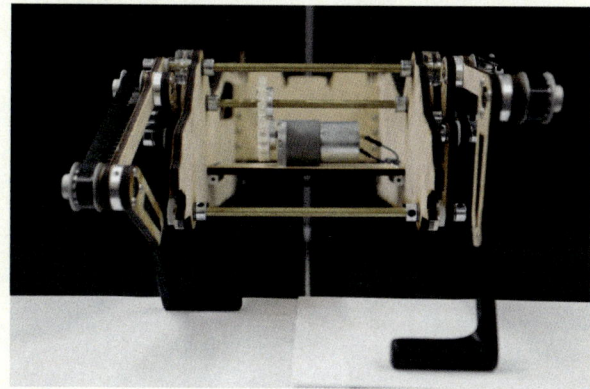

CHAPTER 3

The Two Legged Ostrich

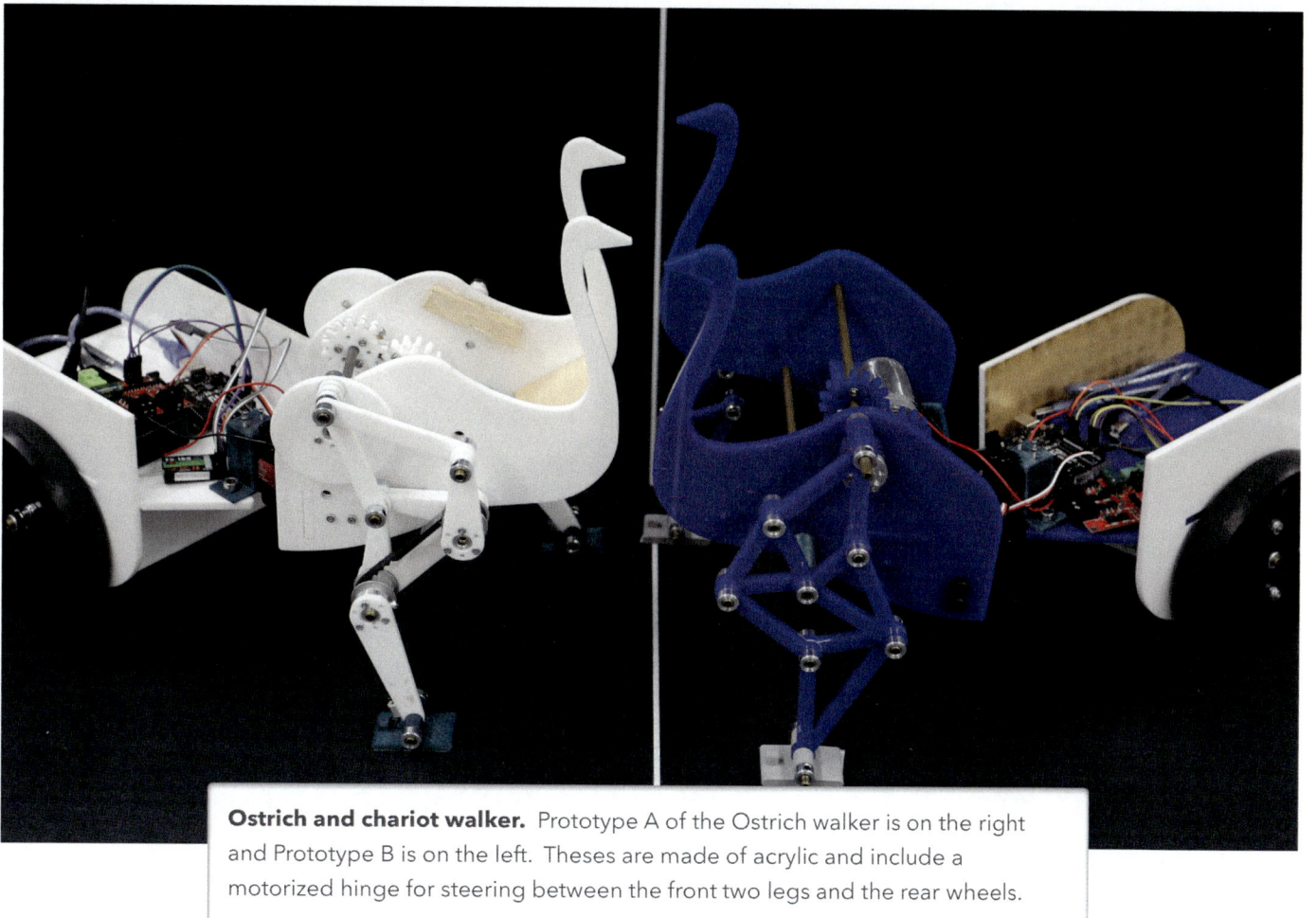

Ostrich and chariot walker. Prototype A of the Ostrich walker is on the right and Prototype B is on the left. Theses are made of acrylic and include a motorized hinge for steering between the front two legs and the rear wheels.

Section 1
Chassis: Acrylic Ostrich Hinged to a Chariot

The two legged Chariot walker was designed to be a steerable mechanical walking robot. The two legged Ostrich in front is connected by a vertical hinge to the rear two wheeled chariot. The vertical hinge is actuated by a servomotor to provide steering.

The shift of loading from one leg to the other during walking introduces alternating torsional loads on the steering hinge that are resisted by the rear wheels.

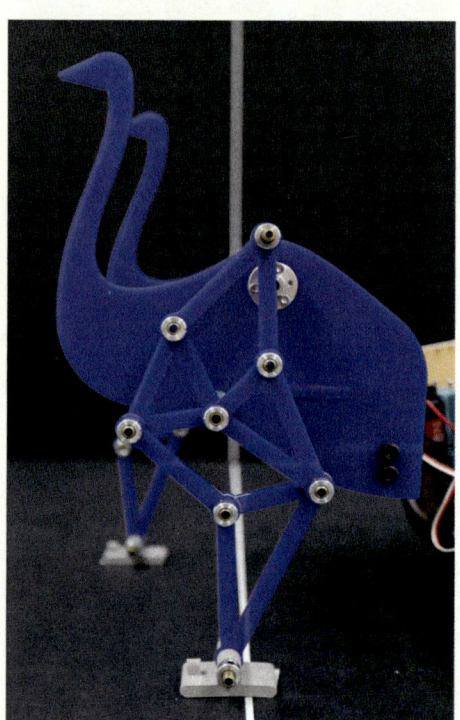

Chassis

The second prototype was constructed address center of mass and leg weakness issues with the first prototype. The chassis for both were constructed from laser cut 1/4 inch acrylic.

Prototype A was designed with a Ostrich that had a track width wider than that of the chariot. This wider track width increased the torsional loads during the shift from one leg to the other which would lift the chariot wheels.

Prototype B reduced the track width of the Ostrich and added cross members to add stiffness to the chassis.

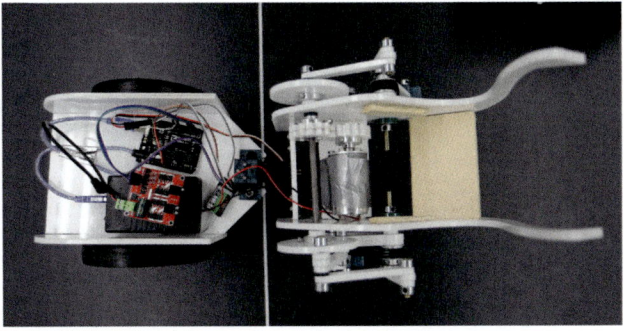

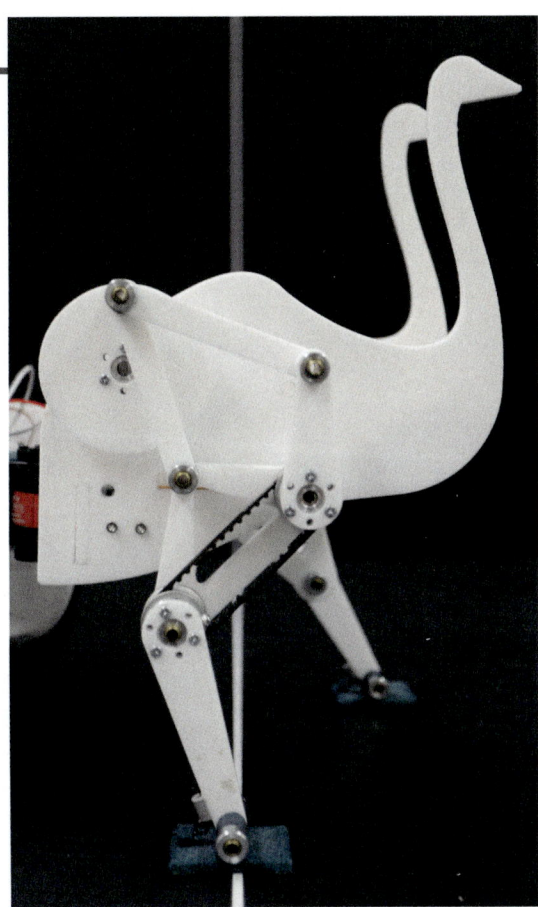

Section 2
Leg Mechanism: Jansen Style Leg

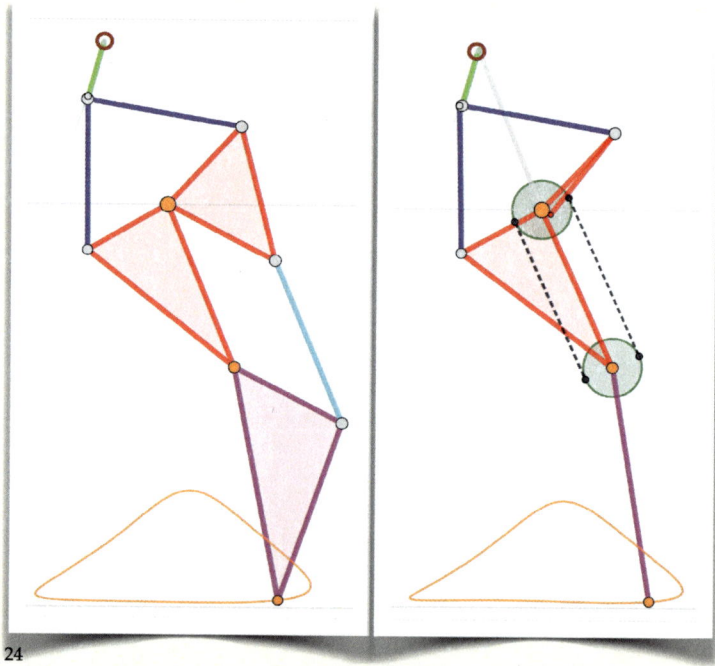

Geogebra provides a convenient tool for the design of a Jansen style leg mechanism. A static model of the topology of the linkage is used to construct a dynamic model, as described in McCarthy (2019). The dimensions of the static model can then be adjusted to achieve a desired foot trajectory in the dynamic model.

The difference between Theo Jansen's leg and our Jansen style leg is the parallelogram that connects the drive link at the hip to the lower leg. This parallelogram decouples the drives of the hip and knee joints.

Furthermore, the parallelogram linkage can be replaced by a belt drive to provide a more reliable leg mechanism.

Digital Prototype

Digital models of Prototype A and Prototype B of the Ostrich walker were used to evaluate the assembly of the systems and their walking movement.

Design studies of the track width and steering hinge position were used to set the dimensions for the two prototypes. These calculations are sensitive to the location of the center of mass.

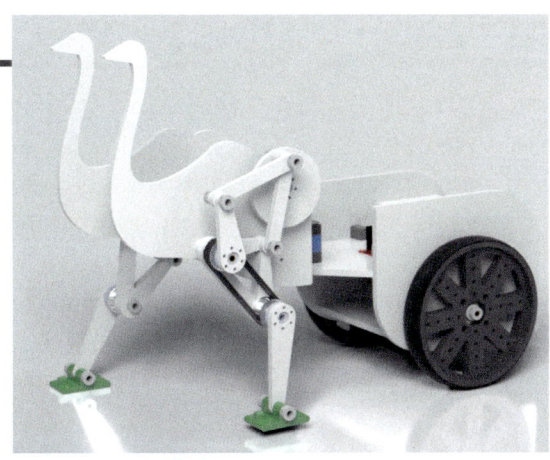

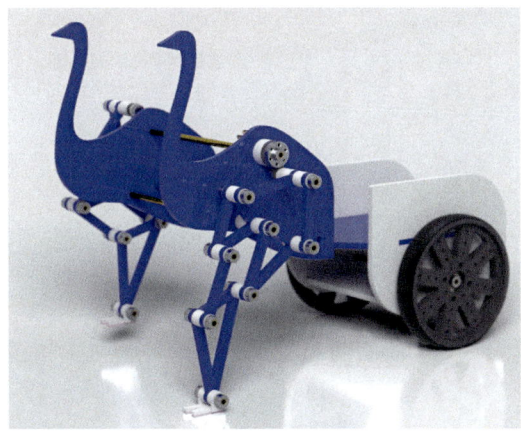

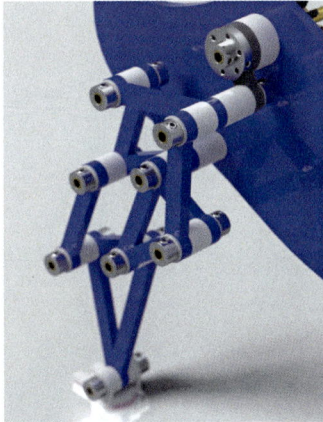

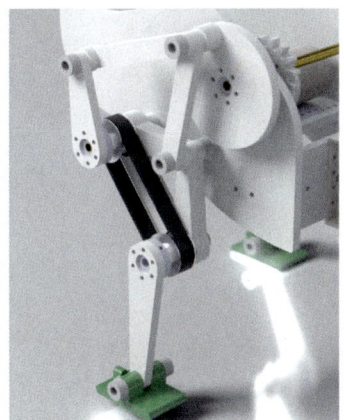

Physical Prototype

Prototype A is constructed using a Jansen style leg, which has a parallelogram linkage that connects the drive link at the hip to the lower leg. It is important to ensure that as the parallelogram linkage does not approach a folding configuration as the leg moves, because internal forces in this configuration can cause the parallelogram to snap through and collapse. It is 45 cm (18 in) long and 32 cm (12.5 in) wide and weighs 2.41 kg.

A timing belt drive was introduced in Prototype B to drive the lower leg instead of the parallelogram linkage. This belt drive performed the same function as the parallelogram, but removes the potential for collapse near folding configurations. This walker is 51 cm (20 in) long and 25 cm (10 in) wide and weighs 2.3 kg.

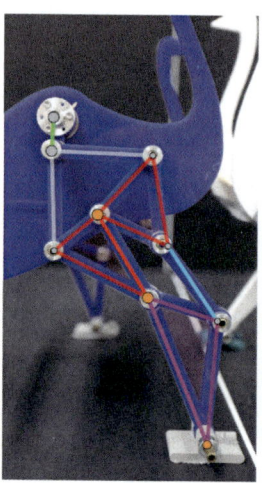

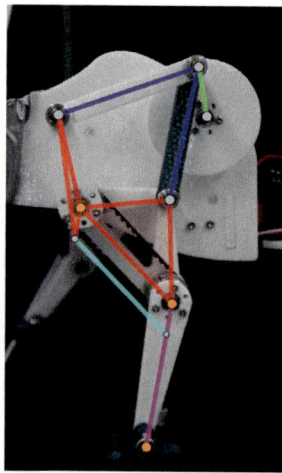
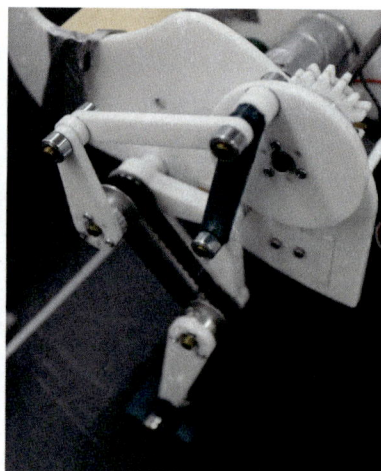

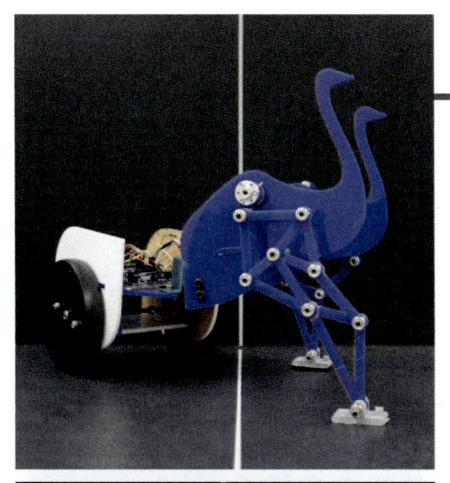
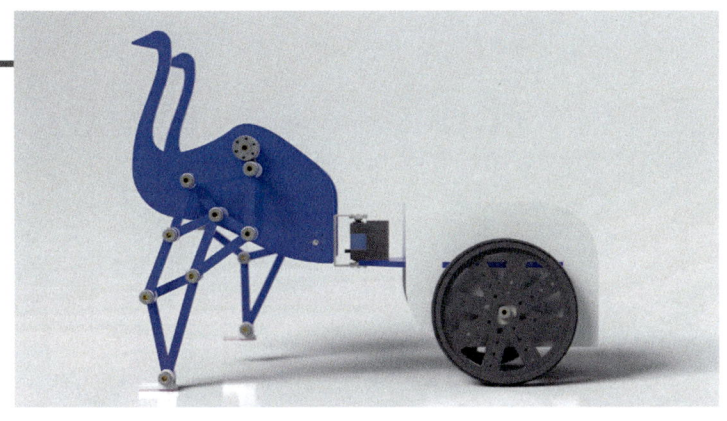
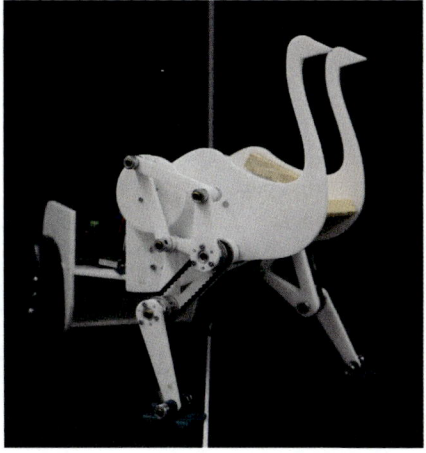
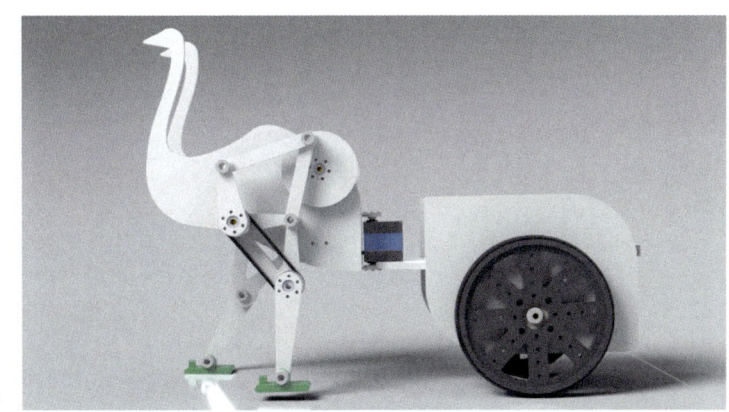

Section 3
Power Train: Two Motors with Acrylic Drive Gears

The Ostrich walker has a drive motor for the leg mechanisms, and a second position controlled servomotor that actuates the central hinge. Input from an RF transmitter actuates the central hinge to provide steering.

Laser cut acrylic gears are mounted on brass tubs using aluminum hubs to drive the leg input cranks.

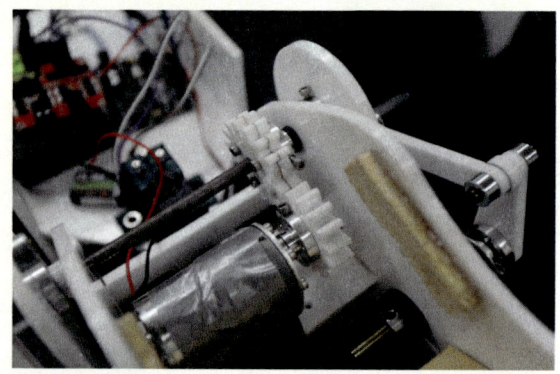

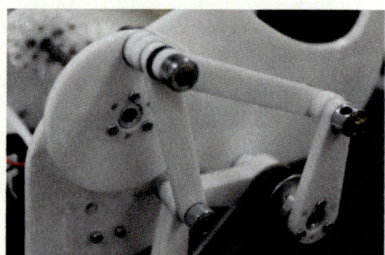

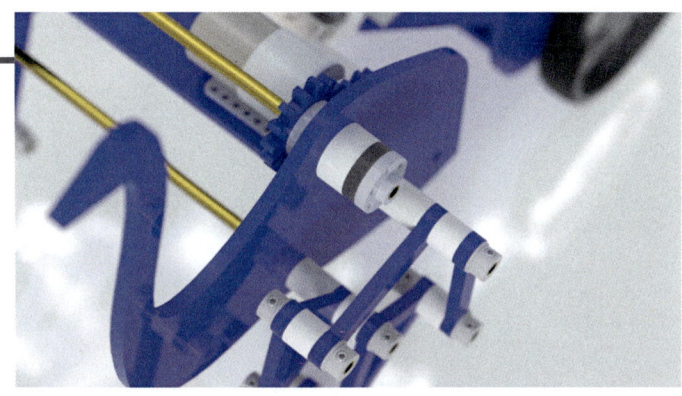
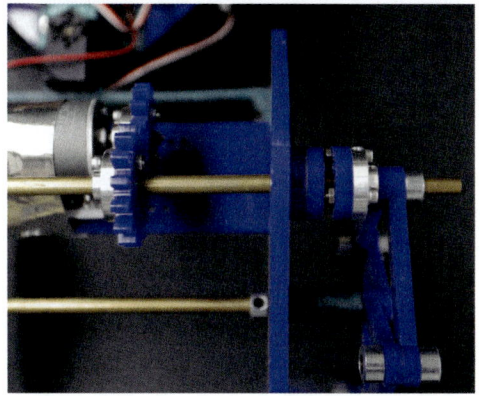
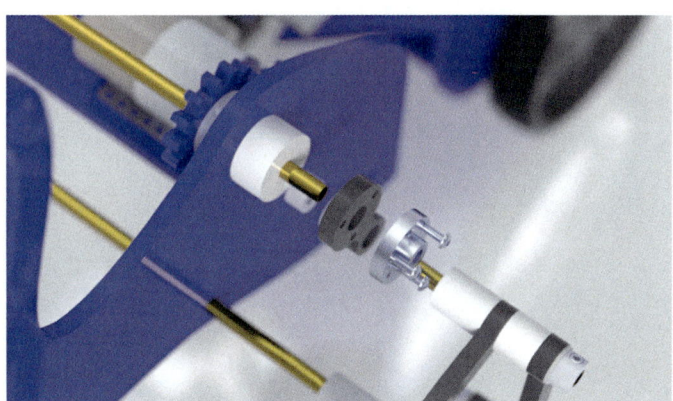

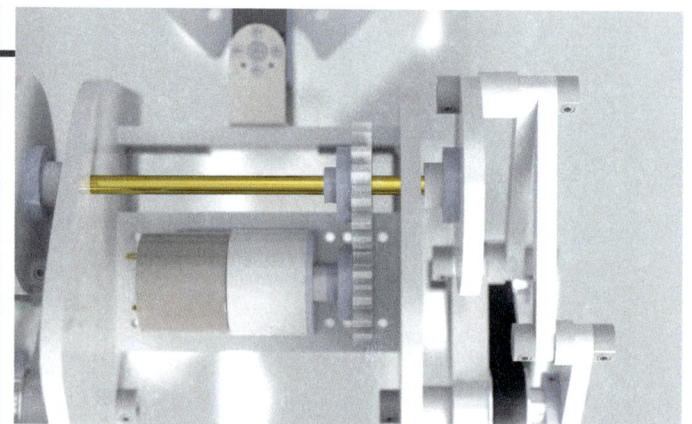

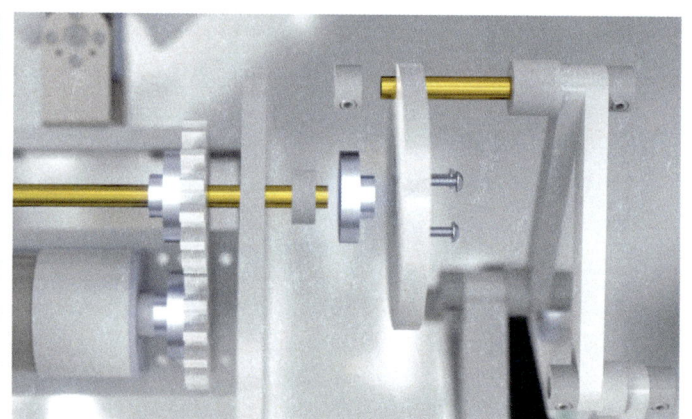

Section 4
Electronics: Speed and Steering Under RF Control

The Ostrich walker implemented speed control of the DC drive motor and steering of the hinge servomotor using a radio-frequency (RF) transmitter often used for RC cars. An RF receiver counted to an Arduino Uno interprets the signal from RF transmitter and sends commands to the motors.

A program on the Uno interprets forward and backward joystick movement is directed to the drive motor and right and leg is directed to the steering servomotor. The schematic for this circuit is shown below.

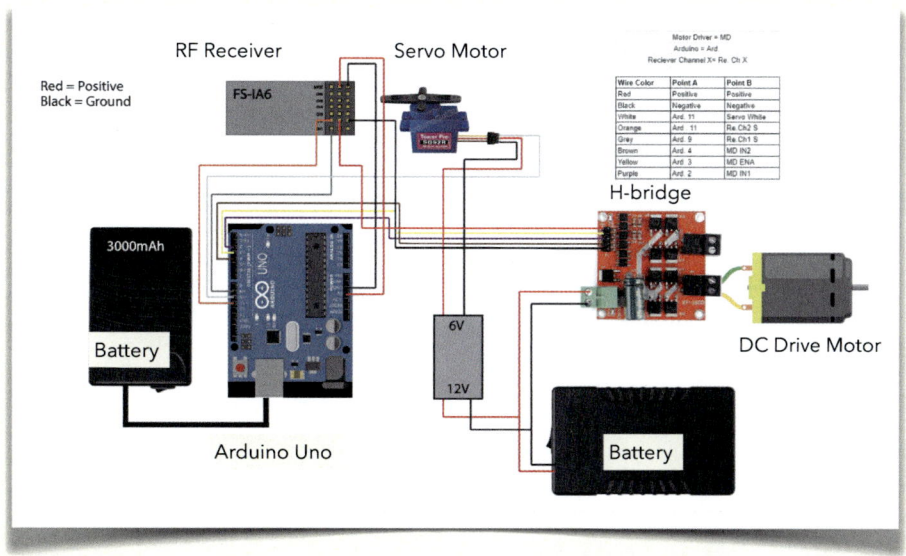

Software

The Arduino program consists of a loop that reads the RC transmitter signal and computes the output to the pulse width modulator (PWM) based on forward and back position of the joystick, which then powers the DC Motor through the H-bridge.

The side to side position of the joystick is read the the RF receiver and signals the servomotor using commands in the servo.h library.

Flowchart: Define variables and parameters and run set-up. → Drive loop → RF Receiver → DC Motor Drive → Steering Motor Drive

```
FINALCODESTEERABLEWALKER §
double channel[2];

//Servo
#include <Servo.h>
int TurnPos;
int val;
Servo ServoAngle;

//DC MOTOR
int enA = 3;
int in1 = 2;
int in2 = 4;
// Motor Speed Values
// |- Start at zero

int joyposVert1;
int joyposVert2;
int joyposVert3;

void setup() {
  Serial.begin(9600);

//Servo
  pinMode(9,INPUT);
  ServoAngle.attach(10);

//DC MOTOR
  pinMode(enA, OUTPUT);
  pinMode(in1, OUTPUT);
  pinMode(in2, OUTPUT);
  pinMode(11,INPUT);
}
```

```
void loop() {
  channel[0] = pulseIn(11,HIGH);
  channel[1] = pulseIn(9,HIGH);
//  Serial.print(channel[0]);
//  Serial.print(" -- ");
//  Serial.println(channel[1]);

  if (channel[0] > 1600)
  {
    digitalWrite(in1, HIGH);
    digitalWrite(in2, LOW);
    joyposVert3 = map(channel[0], 1600, 2000, 0, 255);
    analogWrite(enA, joyposVert3);
  }
  else if (channel[0] < 1400)
  {
    digitalWrite(in1, LOW);
    digitalWrite(in2, HIGH);
    // This produces a negative number
    joyposVert1 = channel[0] - 1500;
    // Make the number positive
    joyposVert2 = joyposVert1 * -1;
    joyposVert2 = map(joyposVert2, 0, 510, 0, 255);
    analogWrite(enA, joyposVert2);
  }
  else
  {
    digitalWrite(in1, LOW);
    digitalWrite(in2, LOW);
    digitalWrite(enA, LOW);
  }
}
```

```
//Steering

if (channel[1] > 1500 && channel[1]< 1400)
{
  val = map(TurnPos, 990,1990,60,125);
  ServoAngle.write(90);
}

else
{ val = map(channel[1],990,2000,60,120);
  ServoAngle.write(val);
}
```

Section 5
Laser Cut Parts

The Ostrich and chariot of both Prototype A and B are constructed from side and cross members laser cut from 12 x 24 inch sheets of 1/4 inch acrylic. The chariots of the two walkers are identical

The leg mechanisms for the two Ostrich walkers are different. Prototype A includes the parts for the parallelogram lower leg drive linkage.

Laser cut plan for the chariots of the Prototypes A and B.

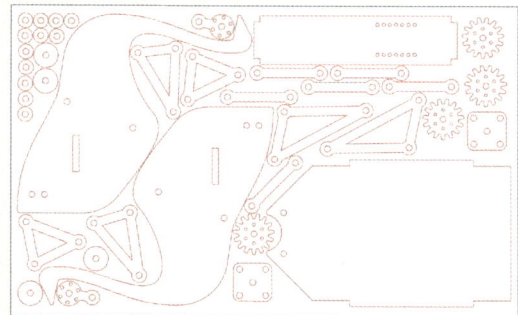

Plan for laser cut of Prototype A components.

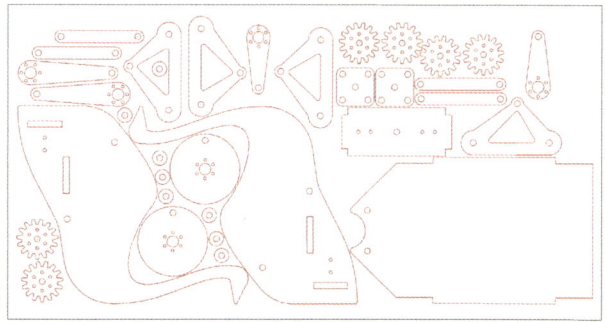

Plan for laser cut of Prototype B components.

Prototype A Parts List

The two walkers are made from laser cut 1/4 in acrylic. The brass tubes move smoothly on acrylic so no bushings or washers are used in the joints. The links, spacers and brass tube are held in place using shaft collars.

The leg mechanisms for Prototype A include the joints of the parallelogram linkages that drive the lower leg portions.

Not included in this list is the breadboard and a number of velcro straps that were used to organize the wiring of the electronics.

Description	Amount		Cost	
Legs, Chassis and Wheels Parts				
6" Heavy Duty Wheel (ServoCity)	2	each	$9.99	$19.98
Cast Acrylic Sheet (White) 12in x 24in x 1/4in	2	each	$21.13	$42.26
Set Screw Shaft Collar for 1/4 in Diameter,	32	each	$1.25	$40.00
Brass Tubing 1/4in OD, 0.032in Wall Thickness	2	3 ft	$7.76	$15.52
Motor Assembly Parts				
50:1 Metal Gearmotor 37Dx54L mm 12V (Pololu)	1	each	$24.95	$24.95
Aluminum L-Bracket for 37D mm Gearmotors (Pololu)	1	each	$7.95	$7.95
Aluminum Mounting Hub for 1/4" (6.35mm) Shaft, M3 Holes (Pololu)	2	2/pkg	$7.95	$15.90
Electronics Assembly Parts				
UCTRONICS 20KG Servo Motor with Bracket Kit.	1	each	$25.99	$25.99
ELEGOO UNO R3 Board ATmega328P	1	each	$13.98	$13.98
Fancasee 5.5mm x 2.1mm Right Angle Pigtail Cord	1	2/pkg	$7.99	$7.99
TalentCell 12V 3000mAh Lithium ion Battery Pack,	1	each	$25.99	$25.99
Kyuionty DC Motor Driver L298 DC Dual H Bridge	1	each	$14.99	$14.99
Flysky FS-i6 6CH RC Transmitter with Receiver	1	each	$49.99	$49.99
Hardware Parts				
Zinc-Plated Machine Screw, M3, 10mm, Phillips	2	25/pkg	$1.29	$2.58
Zinc-Plated Machine Screw, M3, 12mm, Phillips	1	25/pkg	$1.29	$1.29
Zinc-Plated Machine Hex Nut, M3	1	25/pkg	$0.99	$0.99
Socket Head Screw M2 x 0.4mm Thread, 10mm Long	1	100/pkg	$6.75	$6.75
Socket Head Screw M6 x 1mm Thread, 12mm Long	1	50/pkg	$6.81	$6.81
Hex Nut Medium-Strength, M6 x 1 mm Thread	1	100/pkg	$2.59	$2.59
Socket Head Screw 1/4in-20 Thread, 1in Long	1	25/pkg	$6.37	$6.37
Hex Nut, 1/4in-20 Thread	1	100/pkg	$4.88	$4.88
Total				$337.75

Prototype B Parts List

The list of parts for Prototype B differs from the list for Prototype A only in the joints for the parallelogram linkage. This reduced the number of shaft collars from 32 to 24.

The timing belts and pulleys mounted on the upper portion of the each leg mechanism replace the components of the parallelogram linkage.

The belt drive system is costly, but it can increase reliability and performance by avoiding the problems that arise if the parallelogram linkage jams.

Description	Amount		Cost
Legs, Chassis and Wheels Parts			
6" Heavy Duty Wheel (ServoCity)	2	each	$9.99	$19.98
Cast Acrylic Sheet (White) 12in x 24in x 1/4in	2	each	$21.13	$42.26
Set Screw Shaft Collar for 1/4 in Diameter,	24	each	$1.25	$30.00
Brass Tubing 1/4in OD, 0.032in Wall Thickness	2	3 ft	$7.76	$15.52
Motor Assembly Parts			
50:1 Metal Gearmotor 37Dx54L mm 12V (Pololu)	1	each	$24.95	$24.95
Aluminum L-Bracket for 37D mm Gearmotors (Pololu)	1	each	$7.95	$7.95
Mounting Hub for 1/4", 6.35mm Shaft, M3 Holes (Pololu)	2	2/pkg	$7.95	$15.90
Electronics Assembly Parts			
UCTRONICS 20KG Servo Motor with Bracket Kit.	1	each	$25.99	$25.99
ELEGOO UNO R3 Board ATmega328P	1	each	$13.98	$13.98
Fancasee 5.5mm x 2.1mm Right Angle Pigtail Cord	1	2/pkg	$7.99	$7.99
TalentCell 12V 3000mAh Lithium ion Battery Pack,	1	each	$25.99	$25.99
Kyuionty DC Motor Driver L298 DC Dual H Bridge	1	each	$14.99	$14.99
Flysky FS-i6 6CH RC Transmitter with Receiver	1	each	$49.99	$49.99
Hardware Parts			
Zinc-Plated Machine Screw, M3, 10mm, Phillips	2	25/pkg	$1.29	$2.58
Zinc-Plated Machine Screw, M3, 12mm, Phillips	1	25/pkg	$1.29	$1.29
Zinc-Plated Machine Hex Nut, M3	1	25/pkg	$0.99	$0.99
Socket Head Screw M2 x 0.4mm Thread, 10mm Long	1	100/pkg	$6.75	$6.75
Socket Head Screw M6 x 1mm Thread, 12mm Long	1	50/pkg	$6.81	$6.81
Hex Nut Medium-Strength, M6 x 1 mm Thread	1	100/pkg	$2.59	$2.59
Socket Head Screw 1/4in-20 Thread, 1in Long	1	25/pkg	$6.37	$6.37
Hex Nut, 1/4in-20 Thread	1	100/pkg	$4.88	$4.88
Belt Assembly Parts			
XL Series Timing Belt, Trade No. 90xL037	4	each	$9.20	$36.80
Timing Belt Pulley, 3/8" Width with Hub, 1.094" OD	2	each	$5.75	$11.50
Total				$376.05

 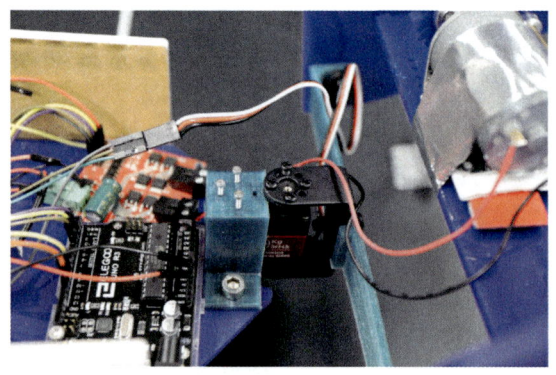

Section 6
Feet

The two walker prototypes were designed to have feet that establish friction contact with the ground. A joint was introduced at the point that traces the flat-sided curve in order to serve as an ankle.

The foot was 3D printed using polycarbonate to interface with the ankle and provide solid support.

Prototype A let the foot float around the ankle as the walker moved along the surface. In Prototype B and small torsion spring was added to keep the foot positions generally horizontally when not in contact with the ground.

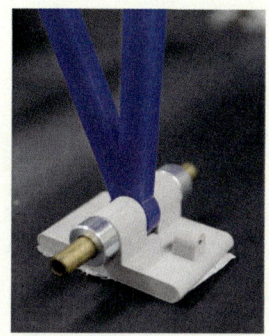
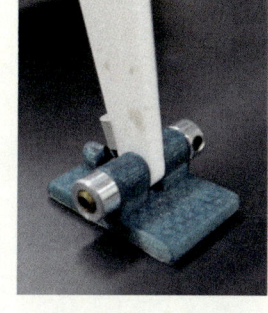
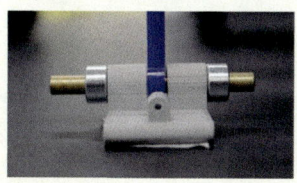

CHAPTER 4

The Four Legged Nightmare

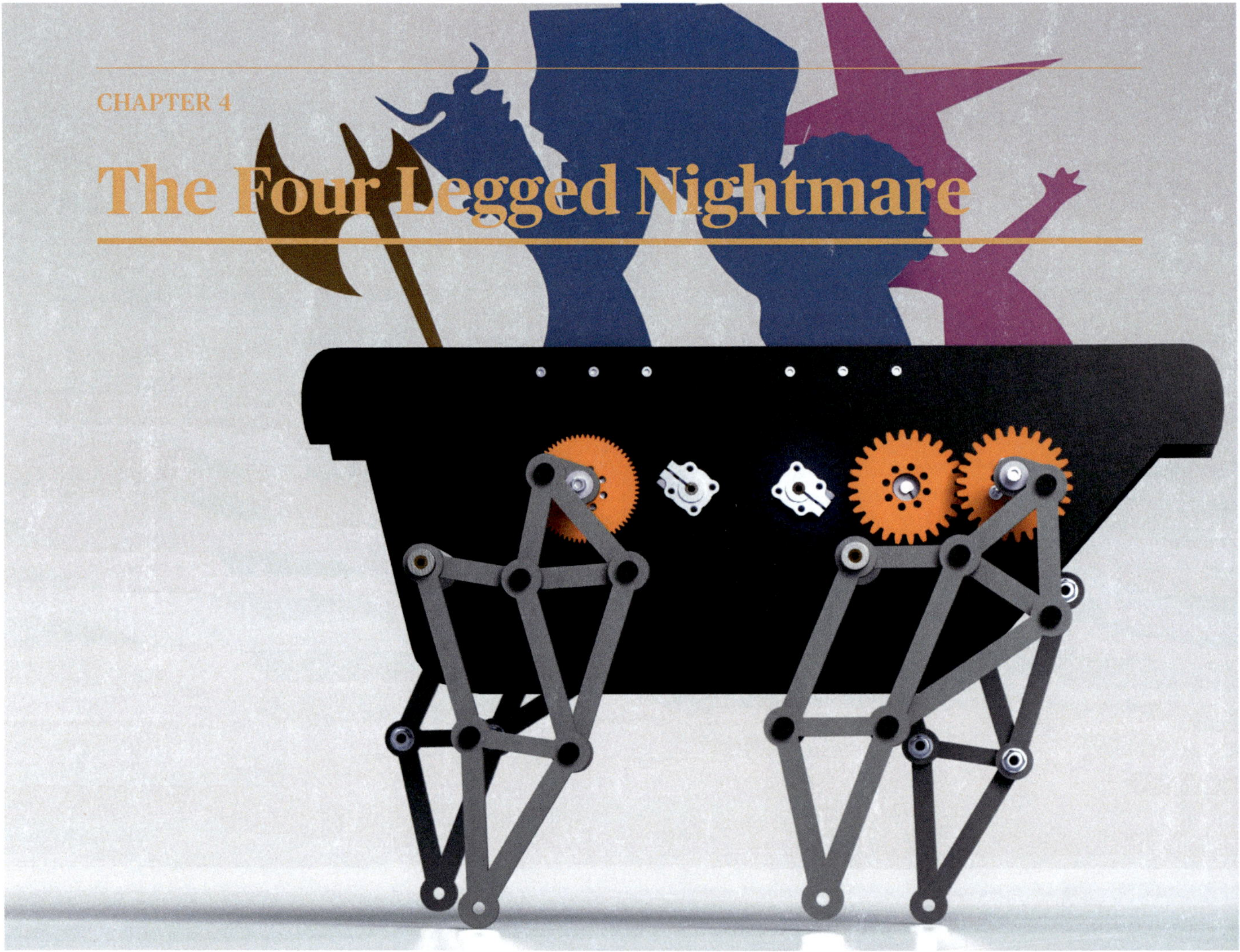

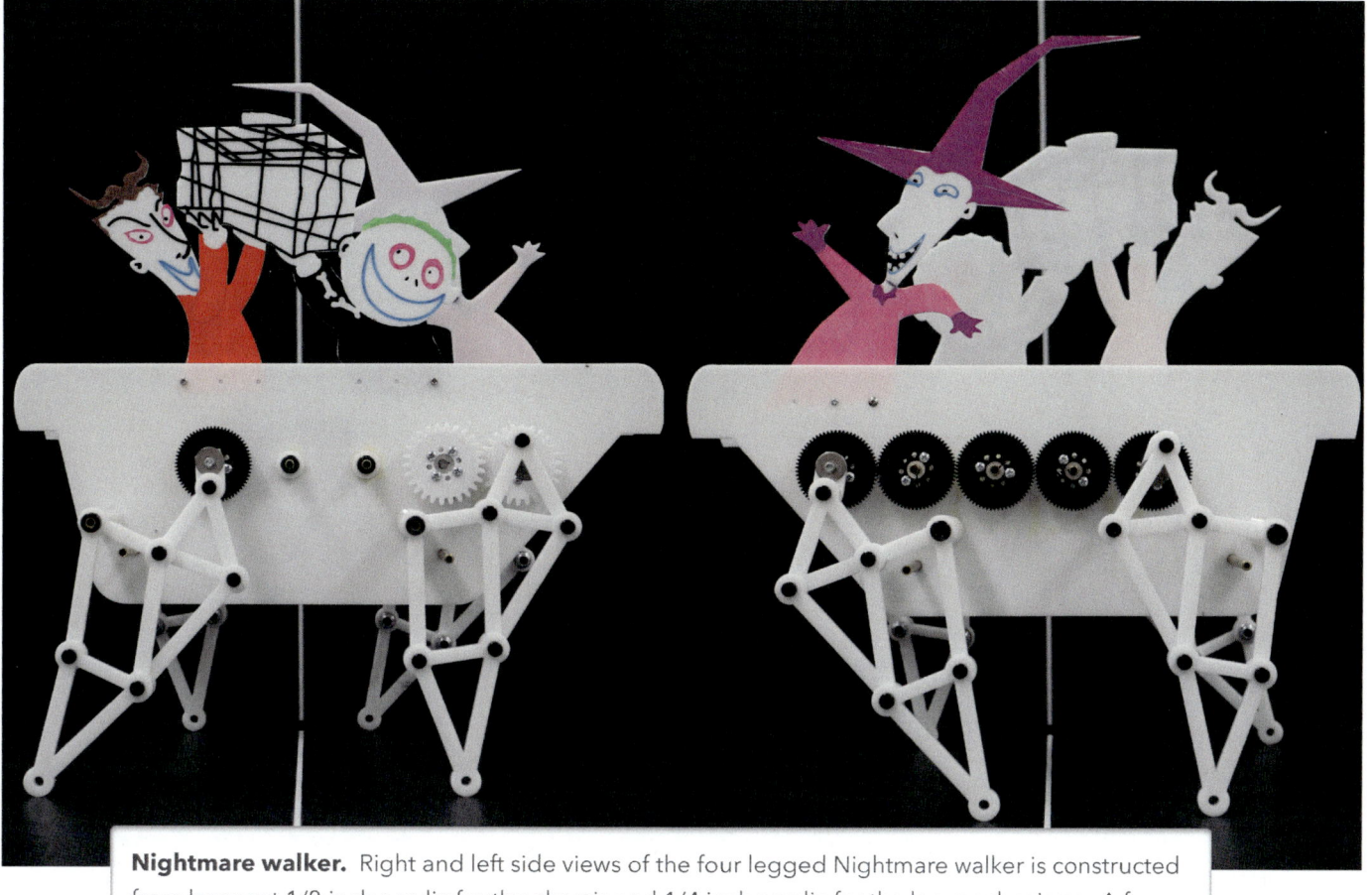

Nightmare walker. Right and left side views of the four legged Nightmare walker is constructed from laser cut 1/8 inch acrylic for the chassis and 1/4 inch acrylic for the leg mechanisms. A four-bar coupler curve is amplified by a skew pantograph to form the leg trajectory.

Section 1
Chassis: Acrylic Bathtub with Nightmare Occupants

The Nightmare walker is made of 1/4 in acrylic and has four legs. The chassis, legs and occupants were laser cut to shape. The occupants were also etched and painted to define their features.

Brass tubes inside 3D printed ABS spacers form the cross members that provide stiffness to the chassis. The acrylic chassis is not as stiff as wood, so the structure flexes as support from the legs shifts from one diagonal to the other.

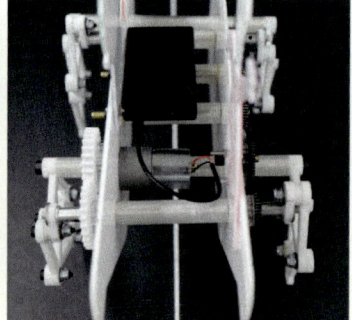

Section 2
Leg Mechanism: Skew Pantograph Leg

The leg mechanism for the Nightmare walker is a skew pantograph six-bar linkage. It is designed starting with a four-bar linkage that has a coupler curve with the desired flat-sided shape.

The skew pantograph is formed by adding a triangle to the coupler link and forming a parallelogram linkage with the output crank. Then the lower leg is constructed to be similar to the triangle added to the coupler link.

The result is the flat-sided coupler curve is scaled and rotated to become the foot trajectory.

The parallelogram linkage of the skew pantograph can be kept from collapsing by adding physical

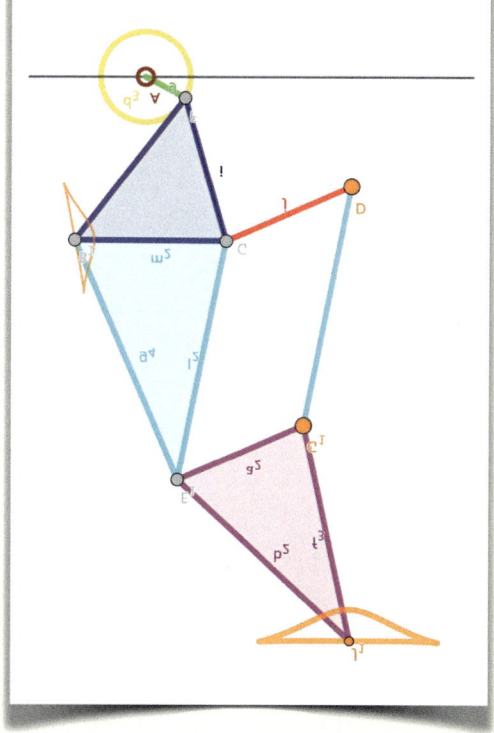

Digital Prototype

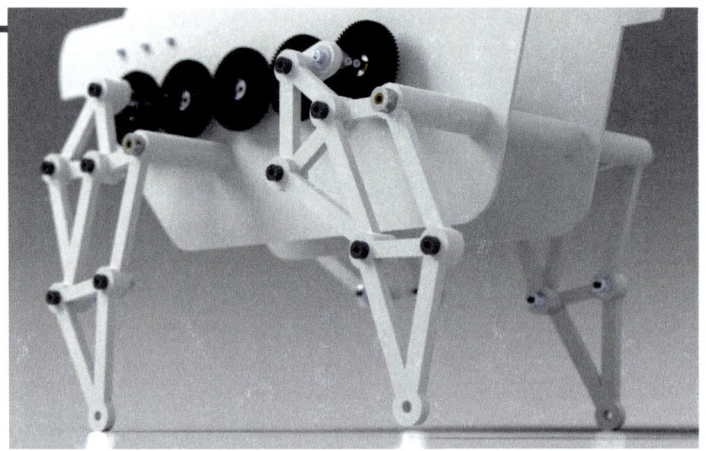

stops at the ends of the range of movement of the lower leg.

The use of wheel mounts and wheel hubs provide a reliable drive for the cranks of the leg mechanisms. The leg joints are formed using nylon shoulder bolts and lock nuts. Acrylic and 3D printed ABS spacers provide the offsets that allow the links to move without interference.

The potential collapse of the leg arising from the near folding position of the parallelogram linkage did not appear until construction of

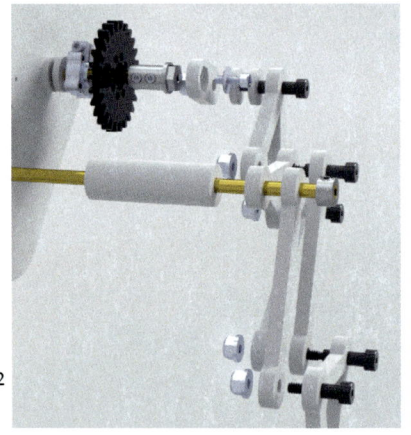
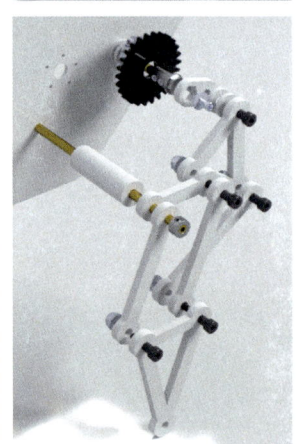

Physical Prototype

the physical prototype. Brass tubes where added to provide hard stops for the movement of the parallelogram linkages so it cannot fold. These tubes also provide increased stiffness for the chassis.

The Nightmare walker is 45 cm (17.5 in) long and 25 cm (10 in) wide and weighs 1.39 kg.

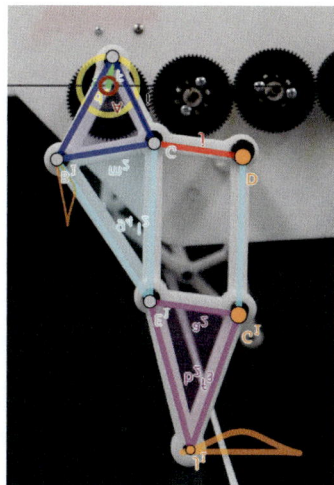

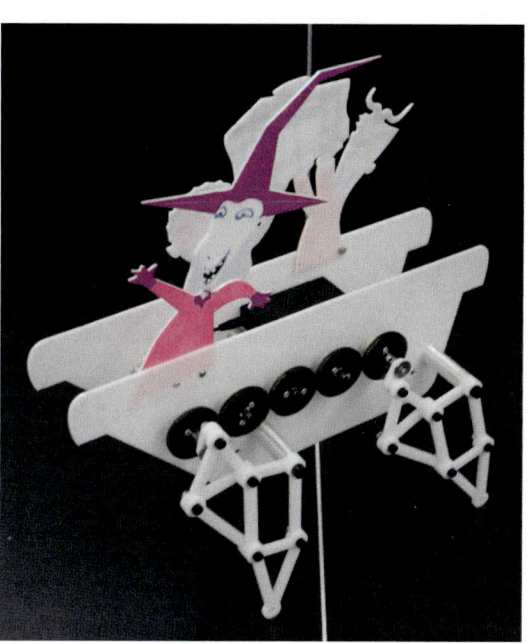

Section 3
Power Train

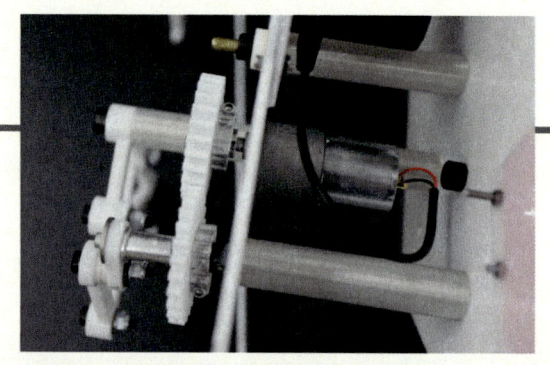

A single motor drives all four legs of the Nightmare walker. 3D printed polycarbonate gears connect the motor to the input crank of the front leg on the right side.

A brass tube connected to the right front drive gear passes through the chassis and rotates the drive gear on the left side. A series of purchased gears down the left side of the chassis drive the rear left leg.

A brass tube connected to the drive of the rear left leg passes back through the chassis to drive the right rear leg.

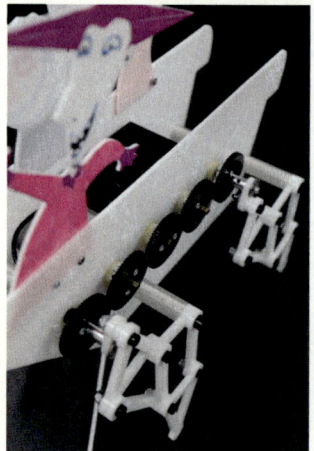
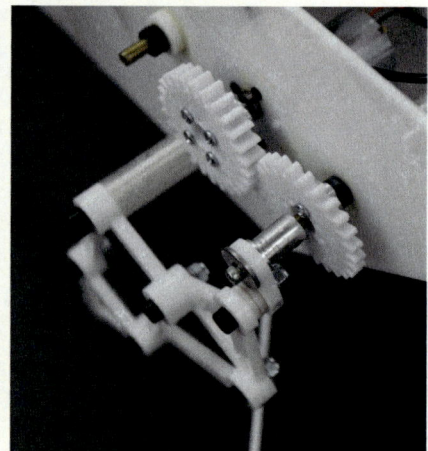

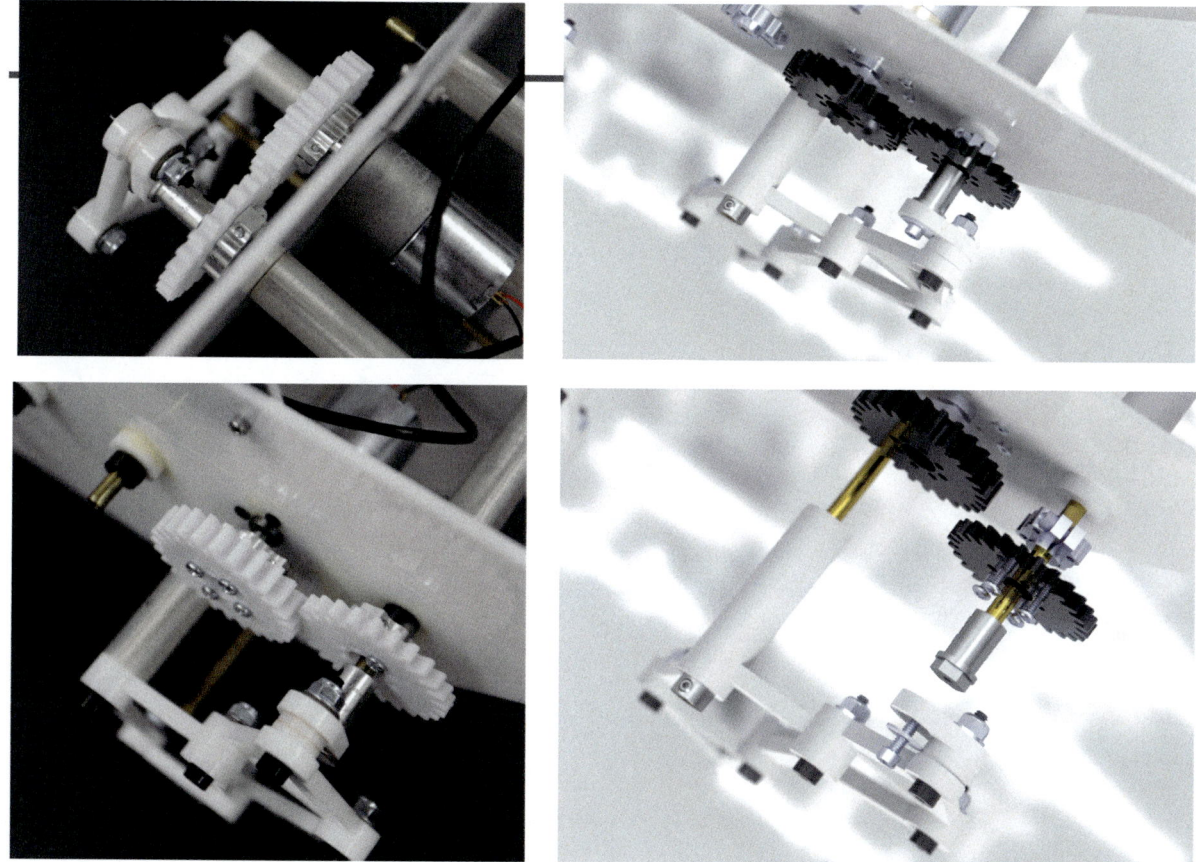

Section 4
Laser Cut Parts

T he bathtub for this walker was obtained by laser cutting 12 inch by 24 inch acrylic sheets 1/4 inch. The three characters and their tools were also laser cut from acrylic and were attached using screws.

The dark lines in the images denote surface etching which does not pass through the acrylic. This is used to create guidelines for painting the characters, as well as to identify the many similar parts in the leg mechanisms.

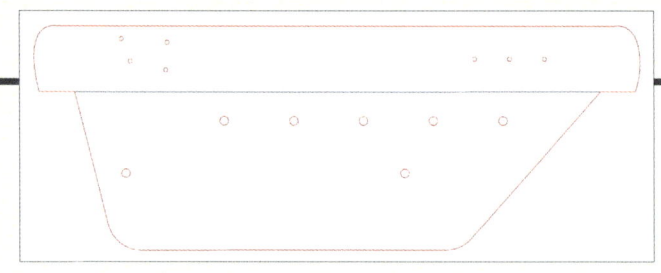

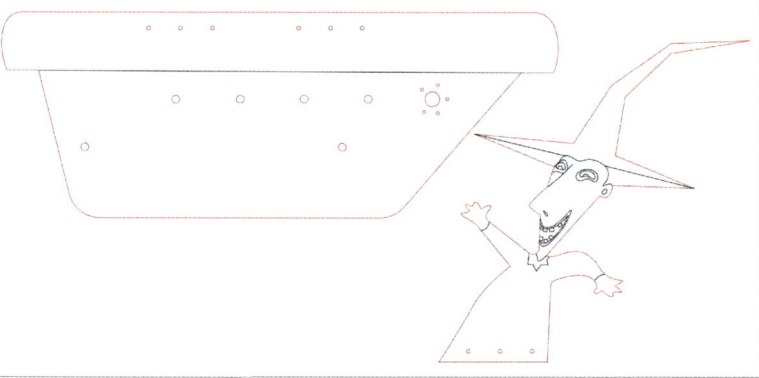

Laser cut plans for the tub chassis and the character Shock.

Laser Cut Parts

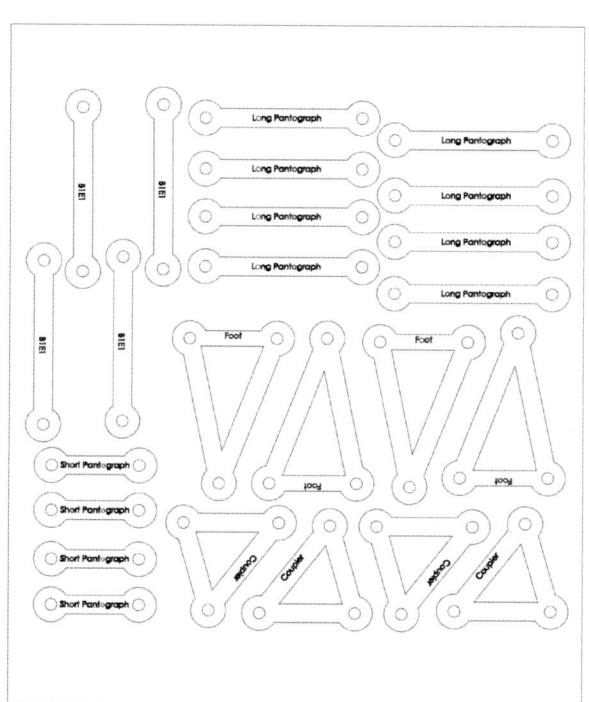

Laser cut plans for the links of the leg mechanisms.

Laser cut plans tools, spacers, and Barrel and Lock.

3D Printed Parts

The Nightmare walker uses 3D printed spacers surrounding brass tubes as structural cross members. The ABS does not print accurately so these spacers must be cut and shaped to size.

Additional structural stiffness was achieved using brass tubes as chassis cross members.

Description	Material	Quantity
Spacer 0.27in ID x 0.75in OD x 2.29in Long	ABS	4
Spacer 0.25in ID x 0.75in OD x 3.35in Long	ABS	6

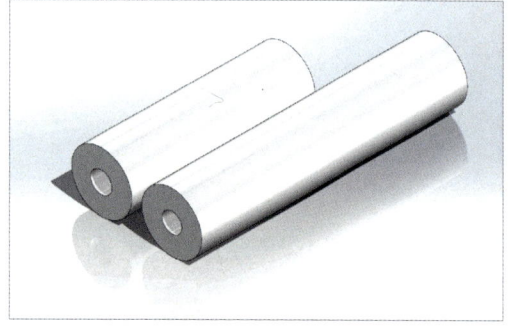

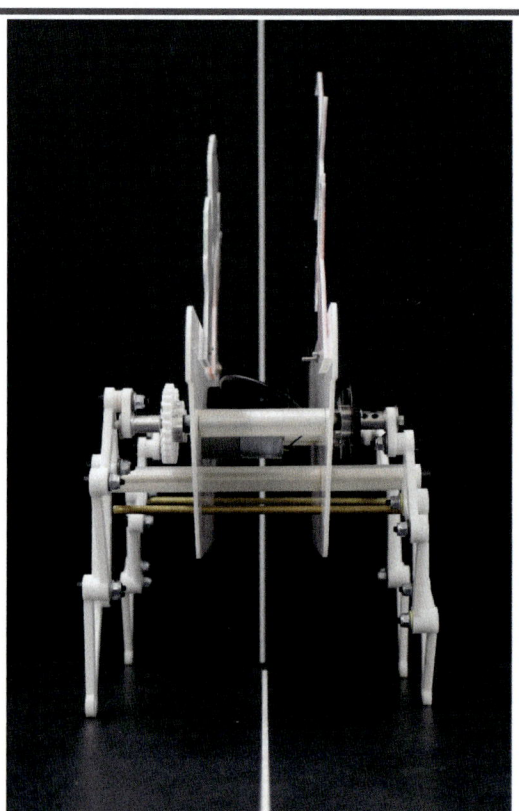

Parts List

The Nightmare walker uses nylon shoulder bolts with steel lock nuts as the joint axes. Hubs are used to connect the motor and brass tube shafts to the gears. And a steel motor mount is used to connect the drive gears to the input cranks for each of the four legs.

Description	Amount		Cost
Leg and Chassis Parts			
Cast Acrylic Sheet 12" x 24" x 1/8"	3	each	$14.27	$42.81
18-8 Stainless Steel Hex Nut 5-40 Thread Size	1	100/pkg	$4.88	$4.88
18-8 Stainless Steel Socket Head Screw 5-40 Thread Size, 7/16" Long	2	50/pkg	$4.20	$8.40
Brass Tubing 1/8" OD, 0.02" Wall Thickness (1ft)	5	each	$4.70	$23.50
Set Screw Shaft Collar for 1/8" Diameter, Black-Oxide 1215 Carbon Steel	20	each	$1.18	$23.60
Medium-Strength Steel Nylon-Insert Flange Locknut Grade F, Zinc-Plated, 10-24 Thread	1	100/pkg	$8.45	$8.45
Nylon Shoulder Screws 1/4" Shoulder Diameter, 1/2" Shoulder Length, 10-24 Thread	1	25/pkg	$8.69	$8.69
Nylon Shoulder Screws 1/4" Shoulder Diameter, 3/4" Shoulder Length, 10-24 Thread	1	25/pkg	$8.44	$8.44
Motor Assembly Parts			
1/4" Bore, 12mm Hex Wheel Mount	2	2/pkg	$7.99	$15.98
50:1 Metal Gearmotor 37Dx54L mm 12V	1	each	$24.95	$24.95
32 Pitch, 64 Tooth Delrin Hub Mount Spur Gear	10	each	$6.16	$61.60
0.125" (0.770") Clamping Hub	10	each	$6.99	$69.90
0.500" (1/2") 6-32 Steel Pan Head Screw	1	25/pkg	$2.19	$2.19
TalentCell 12V 3000mAh Lithium ion Battery Pack	1	each	$25.99	$25.99
Total				$329.38

CHAPTER 5

The Four Legged Chameleon

Chameleon walker. This four legged walker is hinged in the middle to provide steering. The drive motor in the back connects through a drive shaft to the right angle drives in front. A four-bar coupler curve guides a rectangular moving foot to provide the foot trajectory.

Section 1
Chassis: Acrylic Chameleon in Two Sections

The Chameleon walker chassis is constructed of 1/4 in acrylic and consists of two segments connected by a vertical hinge. Each segment has a pair of leg mechanisms, all driven by one motor.

A steering motor actuates the hinge to change the angle between the front and rear chassis segments. This combines with the walking movement to provide steering. The rectilinear leg mechanisms allow the use of large feet which provide stability to the walker.

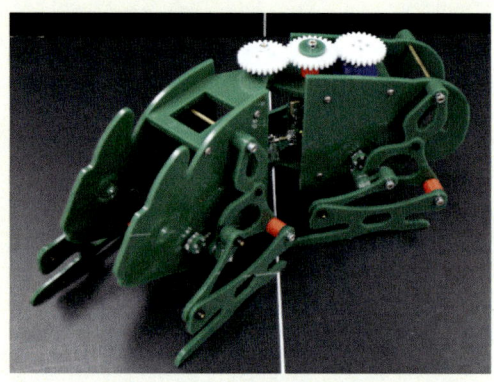

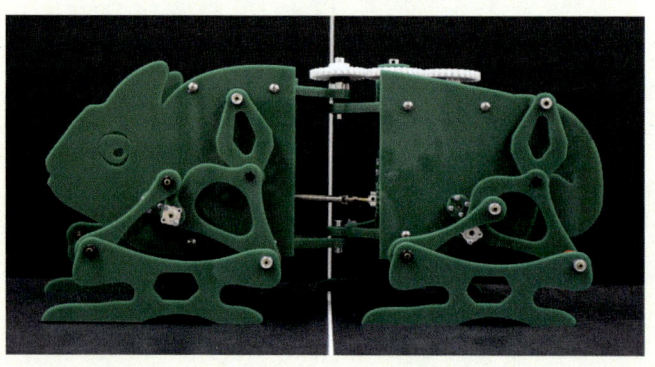

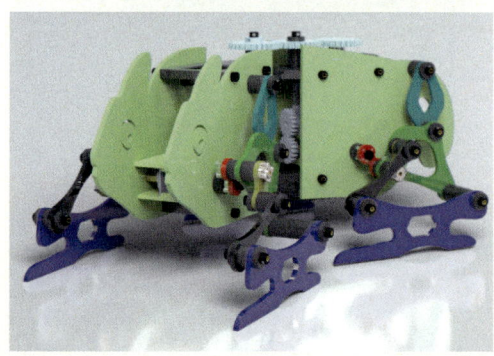

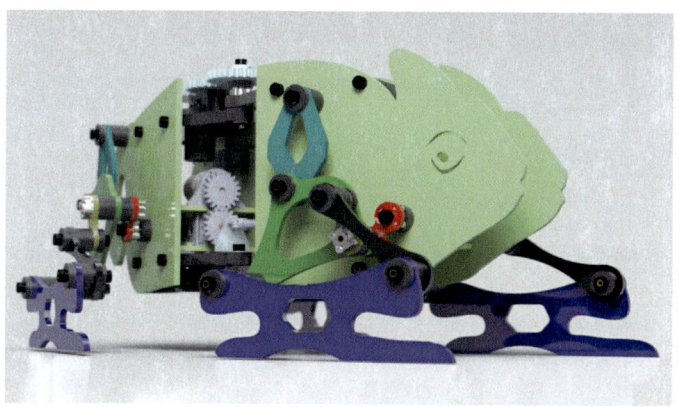

Section 2
Leg Mechanism: Rectilinear Leg

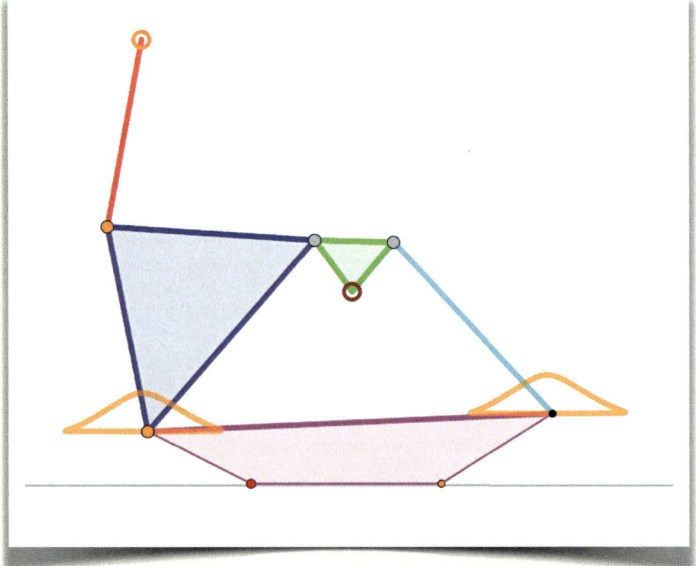

The rectilinear leg mechanism is a six-bar linkage constructed from a four-bar linkage with desired coupler curve foot trajectory. Dijksman (1976) provides a construction that adds two links to a four-bar linkage, one of which traces a specific coupler curve without rotating.

Using this construction, the output link of the four-bar forms the upper leg, the coupler link the lower leg and the rectilinear link becomes the foot. All of the points in the rectilinear link follow the same foot trajectory.

The drive crank for this leg mechanism is a ternary link that connects to both the coupler link and to a second link which connects to the foot. The rotation of the input crank requires the two

Digital Prototype

links it drives to pass over the crank drive pivot without interference. This is achieved providing an offset in the crank that accommodates the movement of these links.

The four legs of the Chameleon are driven by the same motor which is in the rear segment. This motor is aligned with the longitudinal axis of the rear segment and drives the rear legs through a right angle 2:1 bevel gear set. The two input cranks that drive the rear legs are 180 degrees out of phase.

The front legs are connected by the drive motor by a universal drive shaft to a right angle 2:1 bevel gear set. The input cranks to the front legs are connected so they 180 degrees out of phase with the rear legs.

A pair of hinge structures are provided on the top and bottom of the chassis to connect the front and rear segments. The axis of these hinges is connected to 3D printed gears and is driven by a second motor that is oriented vertically in the rear segment.

Physical Prototype

The drive motor connects to the right angle drive for the rear legs through a pair of acrylic spur gears. The input shaft of the right angle drive also connects to the universal drive shaft. This drive shaft connects to the right angle drive of the front legs.

The Chameleon walker is 64 cm (25 in) long and 23 cm (9 in) wide and weighs 4.25 kg.

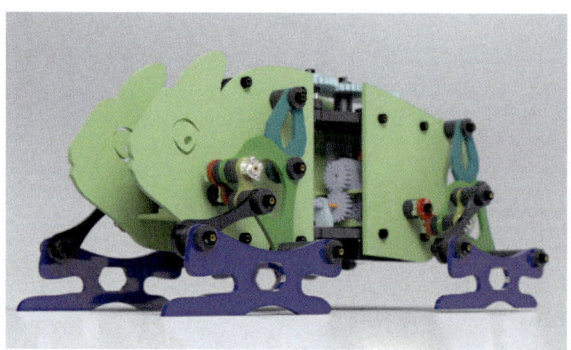
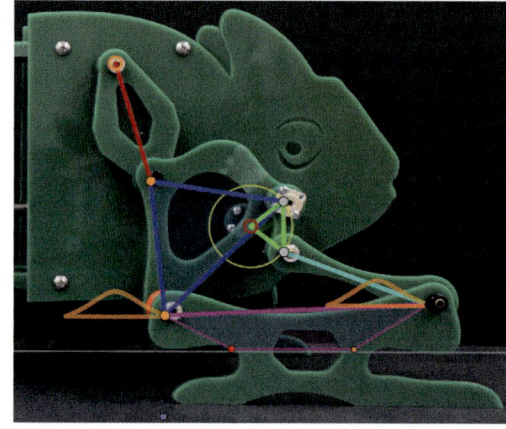

Section 3
Power Train: Drive Motor and Steering Motor

The Chameleon walker has two motors one to drive the legs and a second which provides steering by actuating the vertical axis between the two segments.

The drive motor is connected through a pair of acrylic gears to the input of a right angle drive in the front segment, and through a drive line to the right angle drive in the rear segment.

The cranks on each side of the right angle drive are attached 180 degrees out of phase to provide the alternating diagonal support of the walker.

The steering motor drive two polycarbonate gears that change the angle between the two segments around the vertical hinge. This changes the directions of the leg movement in the two segments and causes the walker to turn.

The four legs of the Chameleon walker are driven by a single motor and coordinated so that the front right and rear left legs have the same movement and the front left and rear right are 180 degrees out of phase. This provides support of the walker along alternating diagonal. The large feet of the walker provide a support polygon that includes the center of mass.

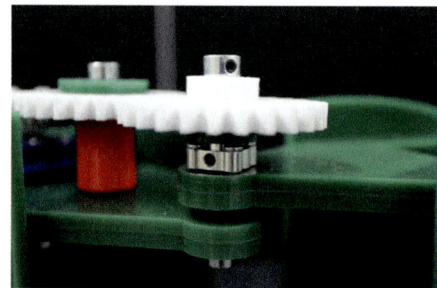

The drive crank of the rectilinear leg mechanism is ternary link that is assembled from two segments that are offset and mounted on hubs to provide a rigid assembly. The lower leg link is attached between the inside and outside segments so it can pass over the axis of the crank. The outside segment connects to the link that supports the foot.

The complexity of this ternary drive crank is required to achieve the benefits of the rectilinear foot movement.

Section 4
Electronics: Speed and Steering Under RF Control

We followed Brandon Tsuge's tutorials on The Bored Robot blog to design the RF system for the speed and steering control of the Chameleon walker.

His schematic and code were adapted to control two DC motor-encoder systems. One for the drive motor and one for the steering motor.

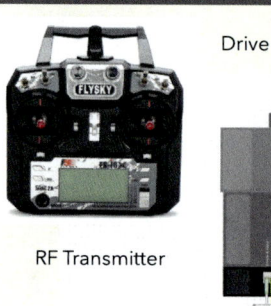

Description	Amount		Cost	
Motors and Battery				
TalentCell 12V 3000mAh Lithium ion Battery	1	each	$25.99	$25.99
100:1 Metal Gearmotor 37Dx57L mm 12V	2	each	$24.95	$49.90
Arduino Nano	1	each	$19.80	$19.80
L298 H-bridge motor driver	1	5/pkg	$10.89	$10.89
Flysky FS-i6X 2.4 GHz RC Transmitter w. Receiver	1	each	$53.99	$53.99
Total				$160.57

Software

This Arduino code to read the RF transmitter and drive the DC drive motor and encoder was developed by Myia Dickens in collaboration with Brandon Tsuge of The Bored Robot blog.

M1 is the drive motor which rotates at constant speed forwards and in reverse. M2 is the steering motor which controls the angle of the vertical steering joint as measured by the encoder.

```
//define motor encoder pins
#define M1EncPin1 2
#define M1EncPin2 3
#define M2EncPin1 4
#define M2EncPin2 5

//define motor encoder counter variables
volatile long Counter1 = 0;
volatile long Counter2 = 0;

//RC PWM variables
unsigned long Ch5PWM = 0;
int tempCh5 = 0;
int Ch5Min = 999;
int Ch5Max = 1987;

unsigned long Ch6PWM = 0;
int tempCh6 = 0;
int Ch6Min = 994;
int Ch6Mid = 1488;
int Ch6Max = 1988;

//Motor output values
unsigned int M1Out = 0;
unsigned int M2Out = 0;
```

```
void setup() {
  Serial.begin(9600);

  //set RC pins to input
  pinMode(Ch5ReadPin, INPUT);

  //set motor driver pins to output
  pinMode(ENA1Pin, OUTPUT);
  pinMode(IN1Pin, OUTPUT);
  pinMode(IN2Pin, OUTPUT);

  //setup to read encoder pins
  pinMode(M1EncPin1, INPUT_PULLUP);
  pinMode(M1EncPin2, INPUT_PULLUP);
  pinMode(M2EncPin1, INPUT_PULLUP);
  pinMode(M2EncPin2, INPUT_PULLUP);
}
```

```
void loop() {
  //read the RC values
  Ch5PWM = pulseIn(Ch5ReadPin,HIGH);
  Ch6PWM = pulseIn(Ch6ReadPin, HIGH);

  //Assign forward/reverse logic for motor controller.
  digitalWrite(IN1Pin,HIGH);
  digitalWrite(IN2Pin,LOW);

  //Calculate Min's and Max's for the RC Channel
  tempCh5 = pulseIn(Ch5ReadPin,HIGH);
  if (tempCh5 < 1000){
    M1Out = 0;
  }
  else{
    M1Out = 255;
  }

  tempCh6 = pulseIn(Ch6ReadPin,HIGH);
  if (tempCh6 < Ch6Min){
    Ch6Min = tempCh6;
  }
  if (tempCh6 > Ch6Min && tempCh6 < Ch6Max){
    Ch6Mid = tempCh6;
  }
  if (tempCh6 > Ch6Max){
    Ch6Max = tempCh6;
  }
  //write values to motor controller
//  analogWrite(ENA1Pin,M1Out);
//  M2Out = map(Ch6PWM, Ch6Min,Ch6Max, -255, 255);
//  analogWrite(ENA1Pin,M2Out);

  Serial.print(tempCh5);
  Serial.println(M1Out);
}
```

Flowchart: Define variables and parameters and run set-up. → Drive loop ↔ RF Receiver, DC Motor Drive, DC Steering Drive

Section 5
Laser Cut Parts

The Chameleon chassis consists of two segments with a vertical hinge. This is the most complex of our walker designs. Four 12 inch by 24 inch acrylic sheets were used to cut the parts.

Nine acrylic gears were cut to connect the drive motor to the four input cranks of the leg mechanisms. Acrylic spacers were used to provide the layering necessary for the six-bar leg mechanisms to move without interference.

A large number of acrylic spacers were used to offset the links of the rectilinear leg mechanism in order to allow the large foot smoothly over the other links. A particular challenge is the ternary link of the drive crank. The movement of another link overlaps the axis of this crank which therefore requires special attention.

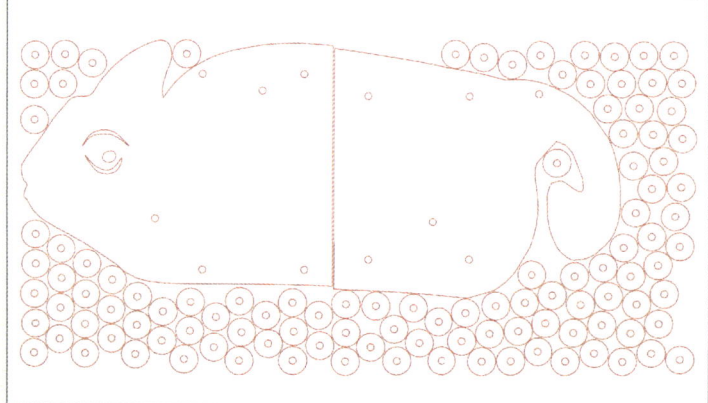

Laser cut plans for the Chameleon chassis, gears and spacers.

Laser Cut Parts

The front and rear chassis segments of the Chameleon walker are held together by a pairs of hinge structures at the top and bottom that form a vertical hinge. Each structure is made from two 1/4 inch acrylic parts to create 1/2 inch thick structures for this hinge.

Brass tubes form the axles of the vertical hinge, which is connected using a hub to 3D printed polycarbonate gears.

Acrylic spacers are used throughout the walker to provide support for the joints while allowing the links to move without interference. The spacers and links are held in place by shaft collars mounted on the brass joint axles.

The acrylic on acrylic and acrylic on brass surfaces have relatively low friction. Silicone grease can be used to reduce this further.

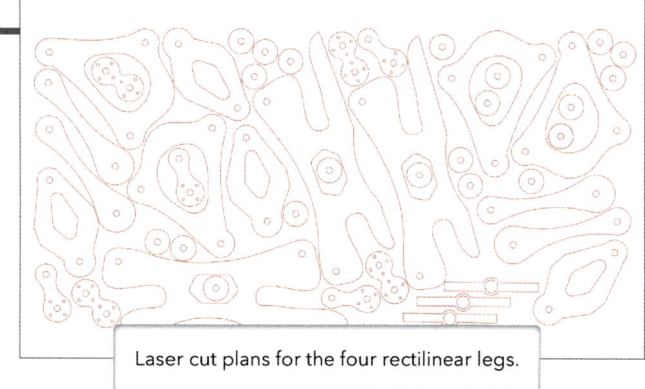
Laser cut plans for the four rectilinear legs.

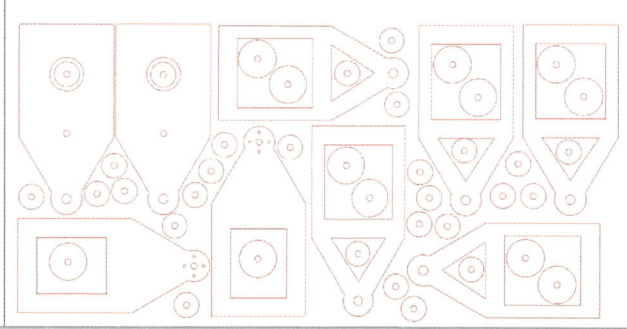
Laser cut plans for the structures that form vertical hinge structures.

3D Printed Parts

3D printed polylactic acid (PLA) spacers were used to provide structural strength to the Chameleon chassis, as well as to create the ternary crank offsets.

The drive gears for the vertical hinge were 3D printed from polycarbonate. Additional 3D printed brackets were used in the motor drive.

An adapter was 3D printed from ABS to connect the universal drive shaft to the drive motor.

Description	Material	Quantity
Hinge gear	Polycarbonate	2
Motor gear	Polycarbonate	1
Adapter, Motor to brass tube	ABS filament	6
Spacer, 0.25 in ID x 1 in OD x 0.875 in Long	PLA	4
Spacer, 0.25 in ID x 1 in OD x 1.17 in Long	PLA	1
Spacer, 0.25 in ID x 1 in OD x 2.5 in Long	PLA	4
Spacer, 0.25 in ID x 0.5 OD x 3.75 in Long	PLA	2
Box Clamp Cap	PLA	2
Box Clamp	PLA	2
Drive Line Motor Mount	PLA	1

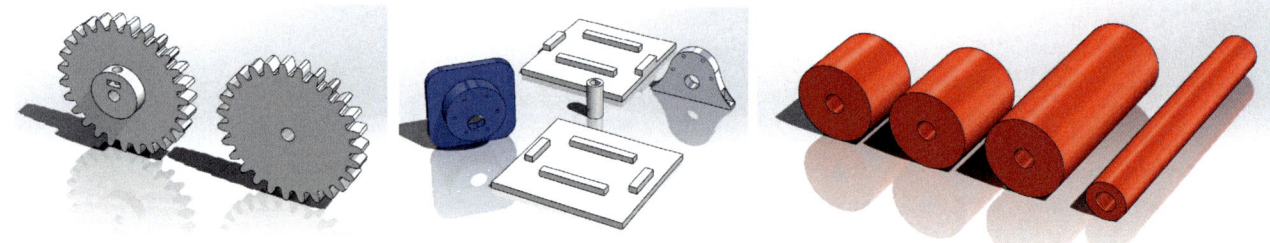

Parts List

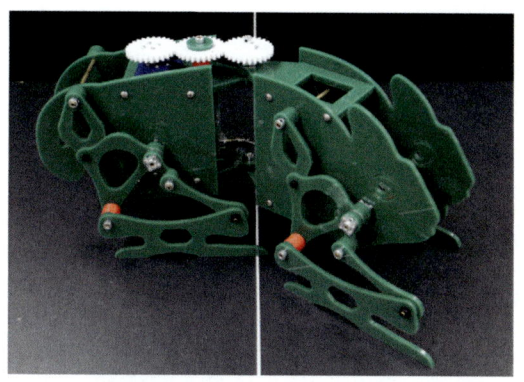

Description	Amount		Cost
Leg and Chassis Parts			
Cast Acrylic Sheet 12" x 24" x 1/4"	4	each	$21.13	$84.52
6" Heavy Duty Wheel	4	each	$9.99	$39.96
Set Screw Shaft Collar for 1/4" Diameter, Zinc-Plated 1215 Carbon Steel	62	each	$1.20	$74.40
Brass Tubing 1/4" OD, 0.032" Wall Thickness	2	6ft	$17.63	$35.26
Dry-Running MDS-Filled Nylon Sleeve Bearing Light Duty, 1/16" Thick Flange, for 1/4" Shaft, 3/8" Long	4	each	$4.36	$17.44
Extra-Wide Truss Head Phillips Screws Passivated 18-8 Stainless Steel, 10-24 Thread, 1/4" Long	1	each	$7.07	$7.07
Drive Line Parts			
Mxfans SCX0016 Universal Drive Shaft	1	2/pkg	$19.11	$19.11
0.250" (1/4) Shafting and Tubing Spacers	1	12/pkg	$1.99	$1.99
1/4" Bore Side Tapped Pillow Block	4	each	$7.49	$29.96
0.250" (1/4") x 2.00" Stainless Steel D-Shafting	2	each	$1.69	$3.38
0.250" (1/4") x 6.00" Stainless Steel D-Shafting	2	each	$3.09	$6.18
U-Channel (3 Hole, 3.00" Length)	2	each	$3.99	$7.98
6-32 x 0.3125" (5/16) Zinc-Plated Socket Head Machine Screw	1	25/pkg	$1.99	$1.99
1/4" ID x 1/2" OD Flanged Ball Bearing	2	2/pkg	$2.89	$5.78
0.250" Bore Clamping D-Hub (Tapped), 0.770" Pattern	2	each	$7.99	$15.98
2:1 Bevel Gear Set (1/4" Bore Pinion, 1/4" Bore Spur)	2	each	$24.99	$49.98
0.250" (0.770") Set Screw Hub	17	each	$5.99	$101.83
0.500" (1/2") 6-32 Steel Pan Head Screw	3	25/pkg	$2.19	$6.57
Slotted Spring Pins 1050-1095, 2mm Diameter, 12mm Long, for 2mm Hole	1	100/pkg	$11.14	$11.14
18-8 Stainless Steel Socket Head Screw 4-40 Thread Size, 3/8" Long	1	100/pkg	$2.81	$2.81
Total				$523.33

Section 6
Feet

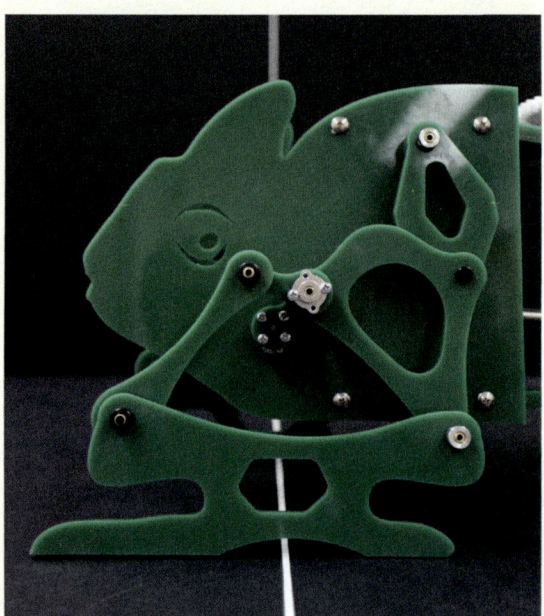

The rectilinear movement of the foot of the Chameleon walker combines a useful foot trajectory with a stable interaction with the ground.

The complexity of the ternary drive crank is balanced by the stability provided by the rectilinear movement of the feet.

This suggests the introduction of rectilinear moving feet for our other leg mechanisms.

CHAPTER 6

The Six Legged Anteater

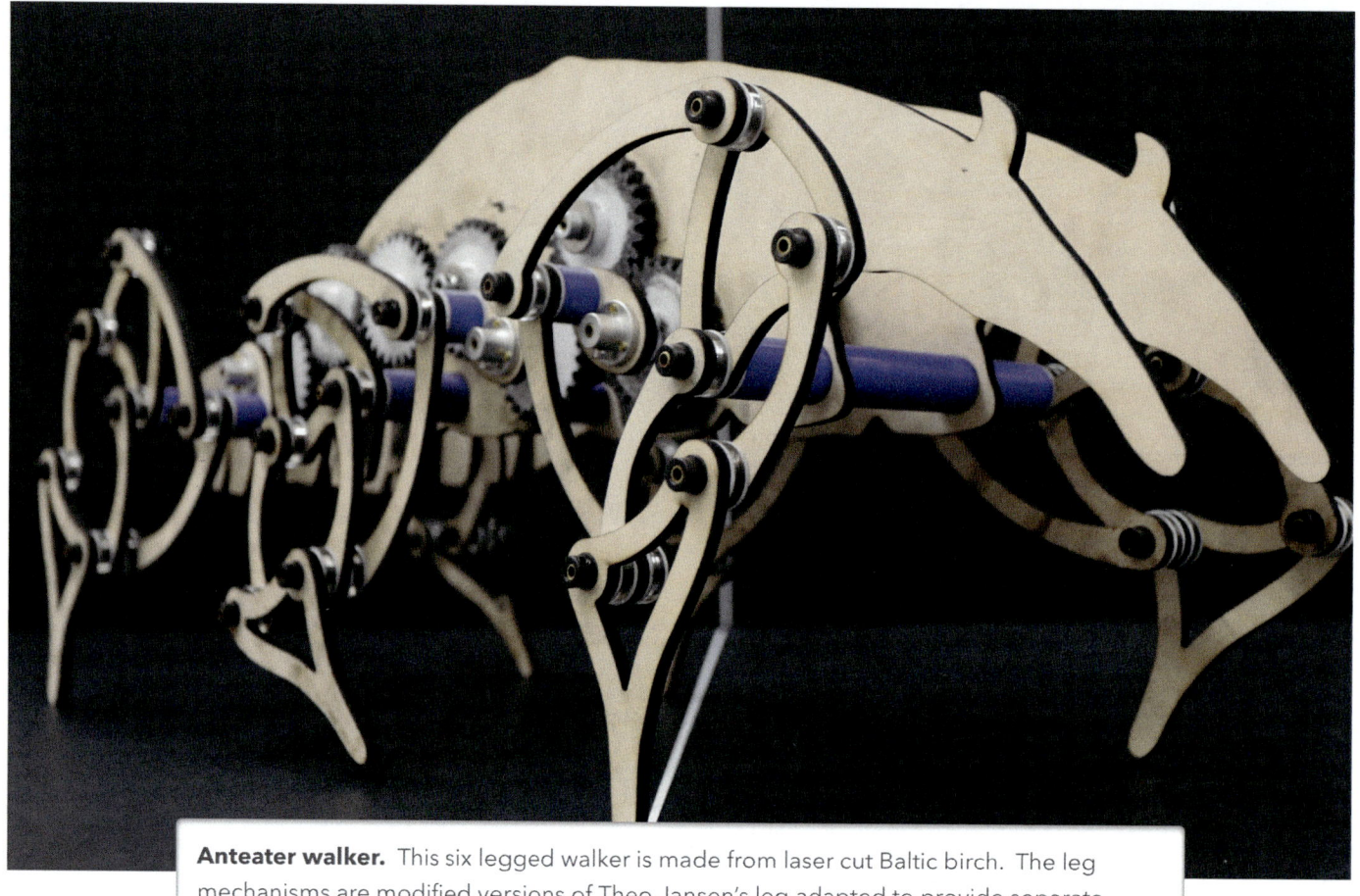

Anteater walker. This six legged walker is made from laser cut Baltic birch. The leg mechanisms are modified versions of Theo Jansen's leg adapted to provide separate control of the hip and knee rotations in order to generate the foot trajectory.

Section 1
Chassis: Baltic Birch Anteater

The Anteater walking robot has six legs that are coordinated so that three at a time are in contact with the ground. This three legged support reduces the twisting forces on the body that arise from intermittent ground contact.

The walker chassis is constructed from two side members cut from 6mm Baltic birch and six 3D printed cross members.

The motor drives a gear train that connects all six leg input cranks. In this walker, there is no computer control, simply an on-off switch on the battery that starts the walking movement.

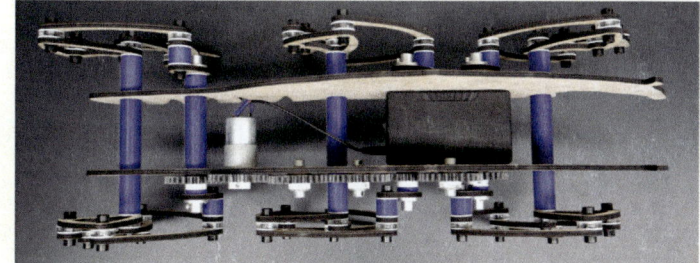

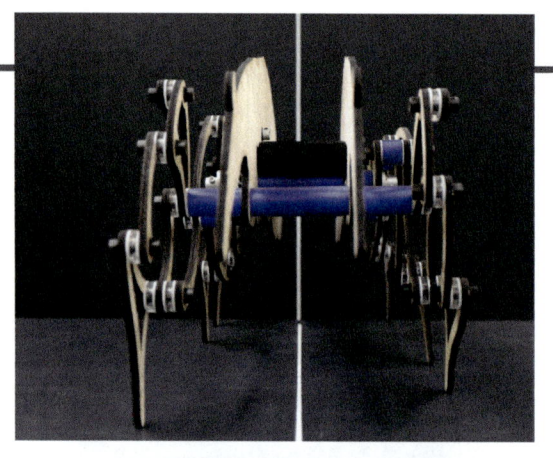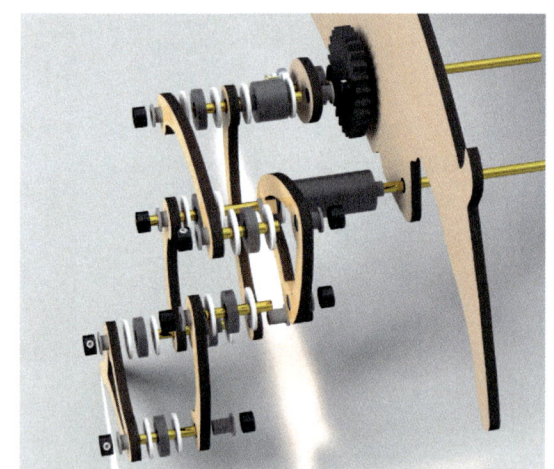

Section 2
Leg Mechanism: Jansen Style Leg

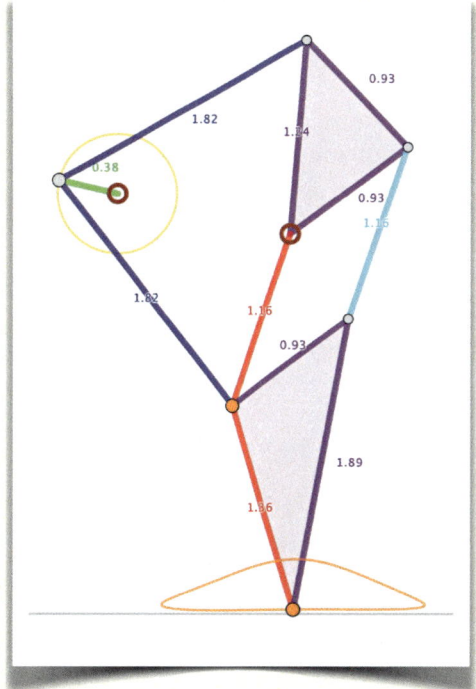

The geogebra model for the leg of the Anteater is a modification of the leg introduced by Theo Jansen. His thirteen "holy numbers" include the near parallelogram with opposite sides given by f=39.4, c=39.3 and d=40.1, g=36.7. By equating these sides so f=c and d=g, we find the rotation of the lower leg about the knee becomes independent of the rotation of the upper leg around the hip.

Adjustment of the other dimensions of this leg yields a symmetric flat-bottomed foot trajectory. Three legs with this trajectory provide a tripod gait for the walker that is smooth and stable.

It is possible to generalize the leg mechanism further by separating the joints that connect three links at the crank and the knee. However, combining joints in this way reduces the number of parts and simplifies construction.

Digital Prototype

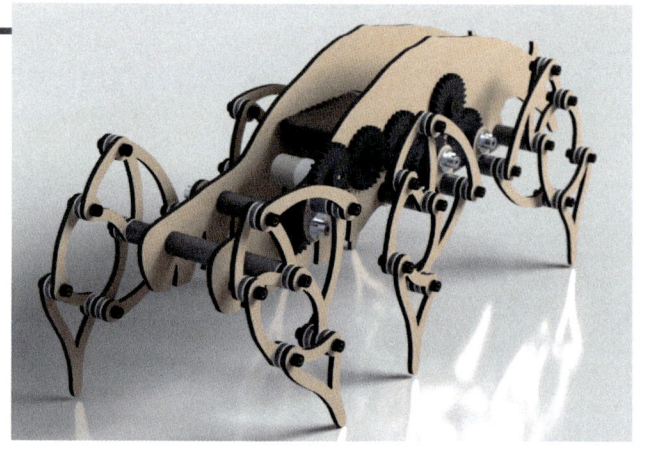

The Anteater leg mechanisms are positioned so the rear two legs are the mirror images of the front two legs. This places the first two legs out of phase by 180 degrees for the same crank rotation.

The natural looking curve of the leg links shows that that they can take any shape as long as they maintain the correct distance between the joints.

The six legs are coordinated so they form a tripod gait, which provides a stable base for the chassis.

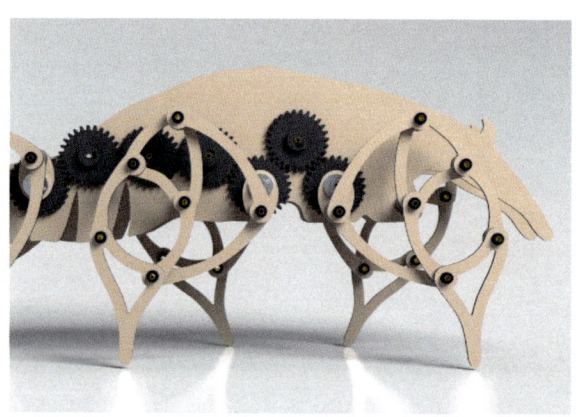

Physical Prototype

The design of the Anteater legs shows that the shape of the links between the joints can take any shape. They simply must provide the strength to maintain the distances required to provide the desired foot trajectory.

The joints for these leg mechanisms are formed from brass tubes with nylon bushings, acrylic spacers and PTFE (polytetrafluoroethylene) washers to provide a low friction surface between the wood of the links and the acrylic spacers, when the shaft collars are squeezed together to ensure the side-to-side strength of the leg at each joint.

A hub is used to ensure an effective connection between the drive gear and the crank of the leg mechanism.

The Anteater walker is 76 cm (30 in) long and 27 cm (10.5 in) wide and weighs 2.95 kg.

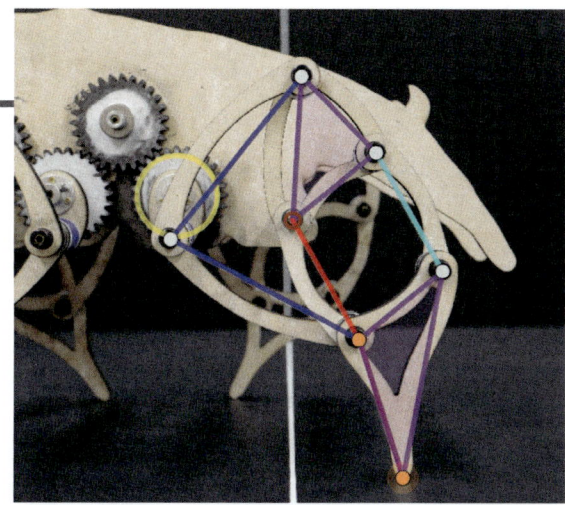

Section 3
Power Train: One Motor with Polycarbonate Gears

A single motor drives the input cranks for each of the six legs. The connection is made through a gear train constructed of 3D printed polycarbonate gears.

Drive cranks on opposite sides of the chassis are connected to the same shaft, so the the drive on one side also drives the other side.

The tripod gait of the six legs is achieved by ensuring the central leg mechanism on each side is 180 degrees out of phase with the front and rear legs; and by setting the two opposite side cranks on each shaft to be 180 degrees out of phase as well.

The alternating supporting tripod formed by the middle leg on one side and the front and rear legs on the other side provides a stable propelling movement for the walker.

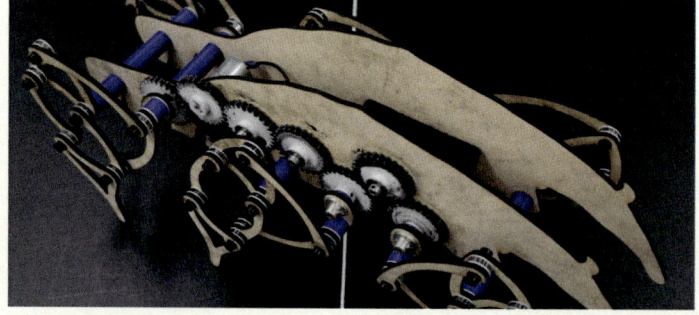

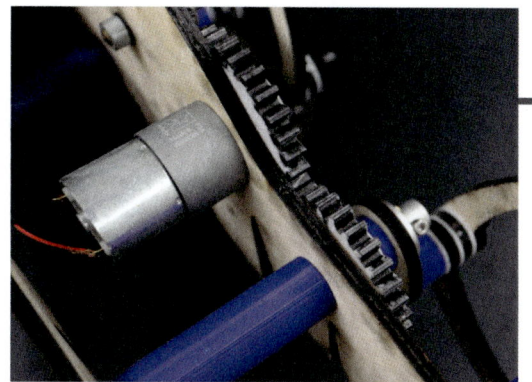

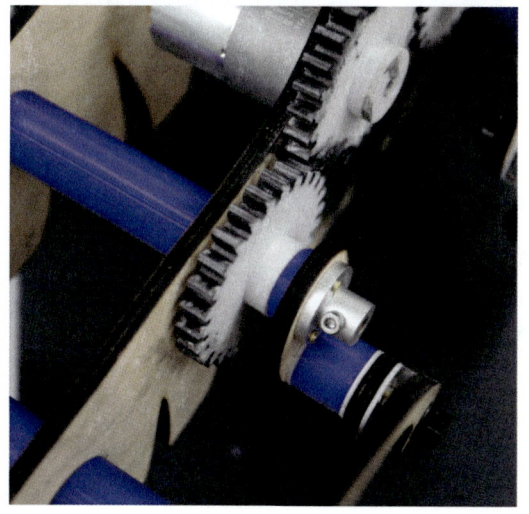

Section 4
Electronics: Infrared Speed Control

O n the Anteater walker we explored the use of an infrared transmitter and receiver to control the speed control of the DC drive motor. An Arduino Uno microprocessor interprets the signal from the infrared transmitter.

The Cytron Shield motor driver mounts onto the Arduino microprocessor and controls the power from the battery to the DC drive motor.

The signal from an infrared remote is obtained by the infrared receiver and interpreted by the Arduino as a motor drive command.

Description	Amount		Cost
Motor, Battery and Electronics			
TalentCell 12V 3000mAh Lithium ion Battery Pack	1	each	$25.99	$25.99
100:1 Metal Gearmotor 37Dx57L mm 12V	2	each	$24.95	$49.90
DAOKI IR Infrared Wireless Remote Control Kits Sensor Board 38KHZ for Arduino AVR PIC	1	each	$4.49	$4.49
ELEGOO UNO R3 Board ATmega328P with USB Cable	1	each	$13.99	$13.99
Cytron Shield-MDD10 10Amp 7V-30V DC Motor Driver	1	each	$20.50	$20.50
Total				$114.87

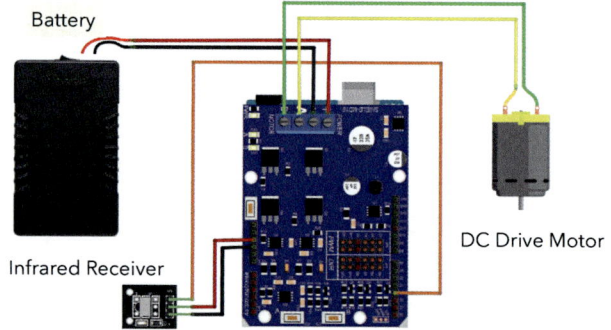

Arduino Uno with Cytron Shield Motor Driver

Software

The infrared remote transmitter provides signals that designate forward and reverse using the up-arrow (24) and down-arrow (82) keys, and half speed forwards and half speed backwards using the right (90) and left arrows (8). The OK key (28) is the stop command.

```
//Functions
//move motors forward at full speed
void Forward() {
  Motor.setSpeed(fullSpeedDirection1);
  Serial.println("Forward");
  return;
}

//move motors backwards at full speed
void Backwards() {
  Motor.setSpeed(fullSpeedDirection2);
  Serial.println("Backwards");
  return;
}

//move motors forward at half speed
void halfForward() {
  Motor.setSpeed(halfSpeedDirection1);
  Serial.println("Forward at half speed");
  return;
}

//move motors backwards at half speed
void halfBackwards() {
  Motor.setSpeed(halfSpeedDirection2);
  Serial.println("Backwards at half speed");
  return;
}

//stop motors
void Stop() {
  Motor.setSpeed(stopMotor);
  Serial.println("Stopped");
  delay(timeDelay);
  return;
}
```

```
void loop() {
  //while the IR sensor is active, listen for a new code
  while (!(IrReceiver.decode()));

  //if data is received, decode it, print to serial monitor
  //then continue to receive new code
  if (IrReceiver.decode()) {

    //decodes only first byte, decodedRawData for all 12 bits
    codeReceived = IrReceiver.decodedIRData.command;
    Serial.print("IR Code:\t");
    Serial.println(codeReceived);

    switch (codeReceived) {
      case 24:
        Stop();
        Forward();
        break;

      case 82:
        Stop();
        Backwards();
        break;

      case 8:
        Stop();
        halfBackwards();
        break;

      case 90:
        Stop();
        halfForward();
        break;

      case 28:
        Stop();
        break;

      default:
        Serial.println("Input not programmed!");
        break;
    }
  }
  //continue to recieve new code
  IrReceiver.resume();
}
```

```
//Cytron Motor Shield Library
#include "CytronMotorDriver.h"

//IR Remote Library
#include <IRremote.h>

//Create motor objects and configure them
//PWM_DIR is defined in the Cytron Library
CytronMD Motor(PWM_DIR, 9, 8);
// Use PWM and Direction control.
// PWM 1 = Pin 9, DIR 1 = Pin 8.)

//Define rpm and direction
//"-" = direction 1, "+" = direction 2
//As defined in the Cytron Library
//0 to 255  = 0-100% Duty-Cycle = Direction 1
//0 to -255 = 0-100% Duty-Cycle = Direction 2
const int stopMotor = 0;
const int fullSpeedDirection1 = 255;
const int halfSpeedDirection1 = 128;
const int fullSpeedDirection2 = -255;
const int halfSpeedDirection2 = -128;
const int timeDelay = 500;

//Sensor data pin
const byte IR_RECEIVE_PIN = 3;

//Define a variable to hold detected IR code
unsigned int codeReceived = 0;

void setup() {

  Serial.begin(9600);

  //Start IR receiver on the arduino
  //pin number and system command DISABLE_LED_FEEDBACK
  //disables led feedback when IR signal is received
  IrReceiver.begin(IR_RECEIVE_PIN, DISABLE_LED_FEEDBACK);
```

Flowchart: Define variables, parameters, and run setup. → Drive loop → DC Motor drive; Infrared Receiver → Drive loop

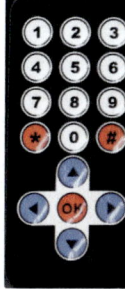

Section 5
Laser Cut Parts

The two sides of the chassis of the Anteater walker was cut from one 12 by 20 inch sheet of 6mm Baltic birch plywood.

A second sheet was used for the leg mechanisms. The rounded shapes of these links provide a more natural look to the walker, avoiding the straight lines connecting the joints that are found in the skeleton drawings.

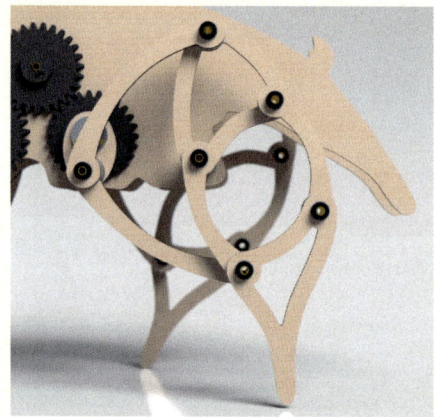

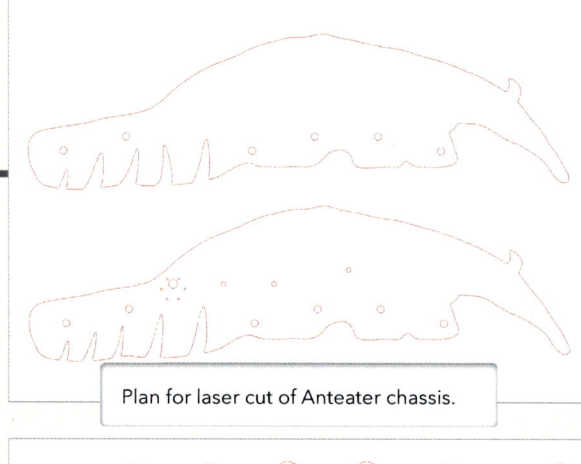

Plan for laser cut of Anteater chassis.

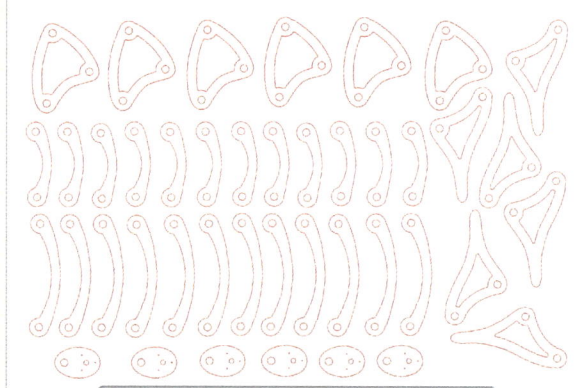

Plan for laser cut of leg components.

3D Printed Parts

Spacers 3D printed out of ABS were designed to fit over the brass tube and nylon bushing assemblies. The outer diameter matches the link dimensions at each joint. The longest of the spacer form the chassis cross members, which yield a stiff chassis structure. The other spacers support the offset amount the links to avoid interference.

The drive gears for the power train were 3D printed from polycarbonate and provide a strong effective transmission.

Description	Material	Quantity
Spacer Crank (0.28in ID x 1in OD x 0.20in Long)	ABS	3
Spacer (0.40in ID x 1in OD x 0.22in Long)	ABS	42
Spacer Outer Crank (0.40in ID x 1in OD x 0.76in Long)	ABS	9
Spacer Leg to Base (0.40in ID x 1in OD x 1.89in Long)	ABS	6
Spacer Body (0.40in ID x 1in OD x 3.40in Long)	ABS	6
Gear (Module = 2.5, Number of Teeth = 27)	Polycarbonate	7

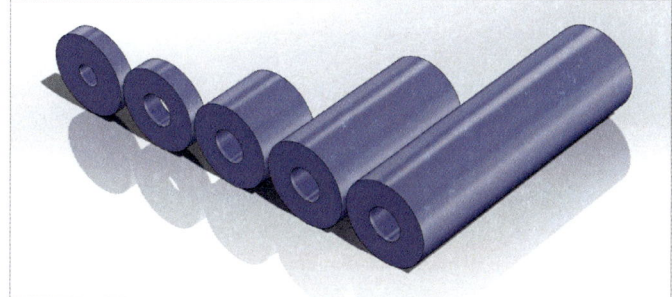

Parts List

The parts for the Anteater walker were selected for their availability through on-line orders with only days required for delivery. And, the intent was that only hand tools would be needed for assembly.

In particular, a hack saw and file was used to size the brass tube and screw drivers were used to assemble the shaft collars, nut strips and hubs. Wiring the motor requires soldering two wires. The electronics was assembled without soldering. Therefore, the parts are relatively expensive but the assembly is easy.

Description	Amount		Cost	
Legs and Chassis Parts				
Brass Tubing 1/4" OD, 0.032" Wall Thickness	3	Each	$11.76	$35.28
Set Screw Shaft Collar for 1/4" Diameter, Black-Oxide 1215 Carbon Steel	75	Each	$1.20	$90.00
Dry-Running MDS-Filled Nylon Sleeve Bearing Light Duty, 3/64" Thick Flange, for 1/4" Shaft, 1/2" Long	50	Each	$1.48	$74.00
Dry-Running MDS-Filled Nylon Sleeve Bearing Light Duty, 3/64" Thick Flange, for 1/4" Shaft, 3/8" Long	30	Each	$1.10	$33.00
Chemical-Resistant PTFE Plastic Washer for 3/8" Screw Size, 0.406" ID, 1" OD	11	10/pkg	$12.99	$142.89
Dry-Running MDS-Filled Nylon Sleeve Bearing Light Duty, for 1/4" Shaft Diameter, 1/2" Long	50	Each	$1.29	$64.50
Motor Mount Parts				
6mm Aluminum Mounting Hub for 60mm Mecanum Wheel	6	Each	$6.21	$37.26
Ultra-Low-Profile Socket Head Screw Alloy Steel, M3 x 0.50 mm Thread, 8 mm Long	6	Each	$3.54	$21.24
Total				$498.17

Walking

This walker does not have feet, it uses friction between the lower leg and the ground to generate propelling forces. The tripod gait provides a stable support that removes twisting forces on the feet that occur in two legged and four legged walkers.

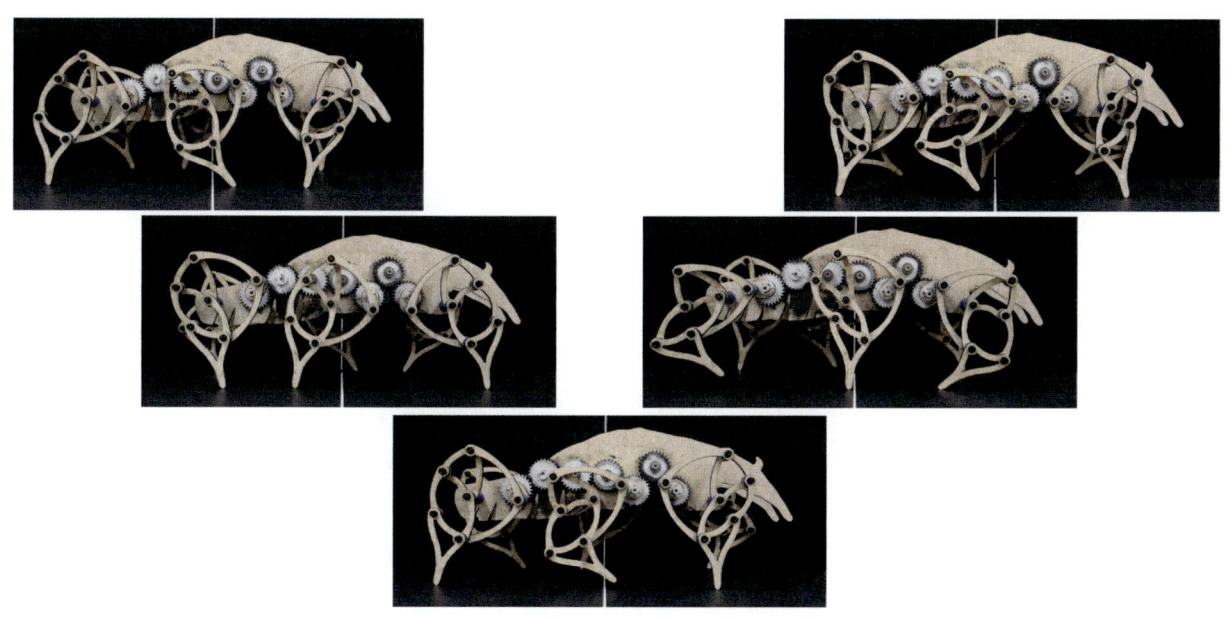

CHAPTER 7

Conclusion

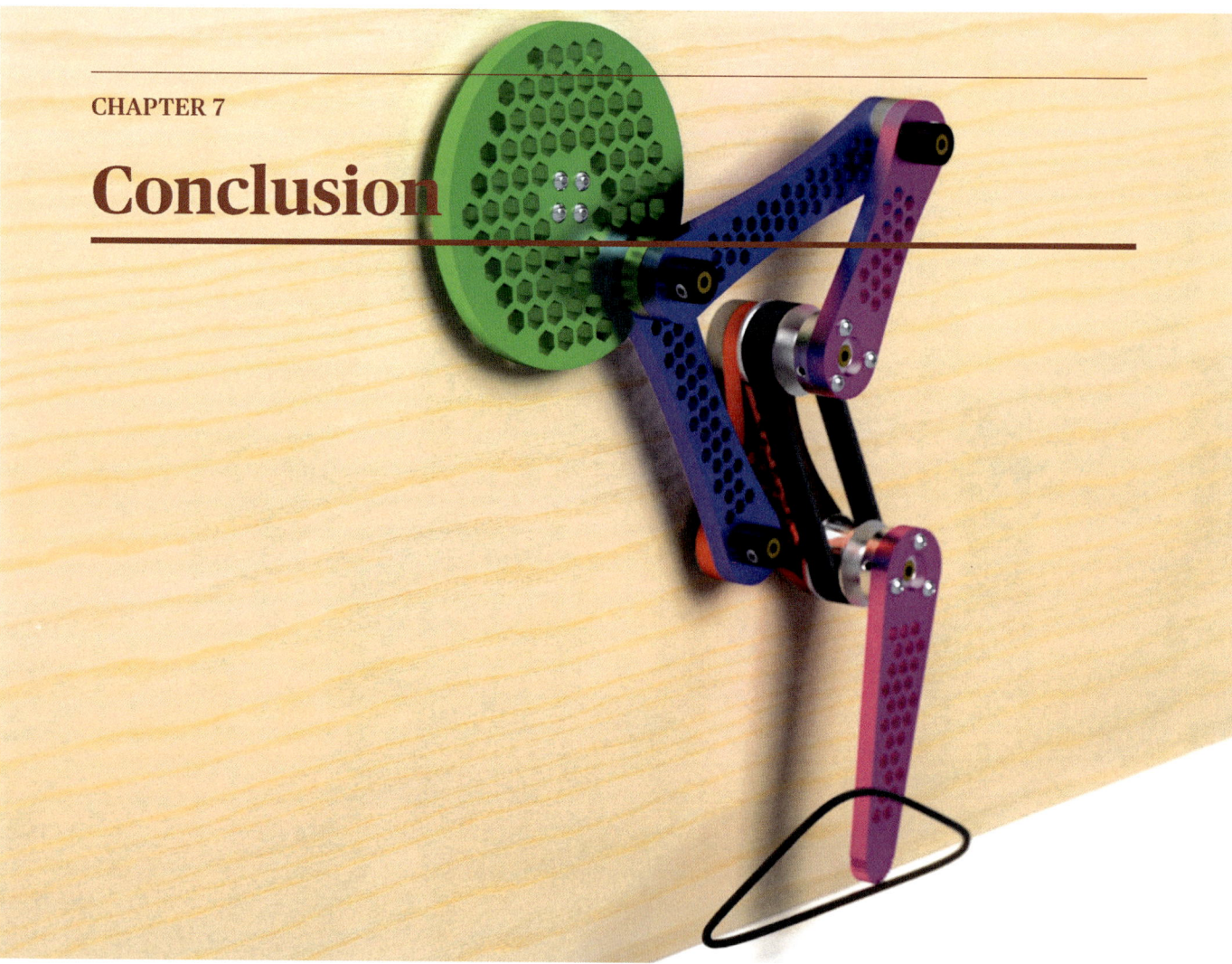

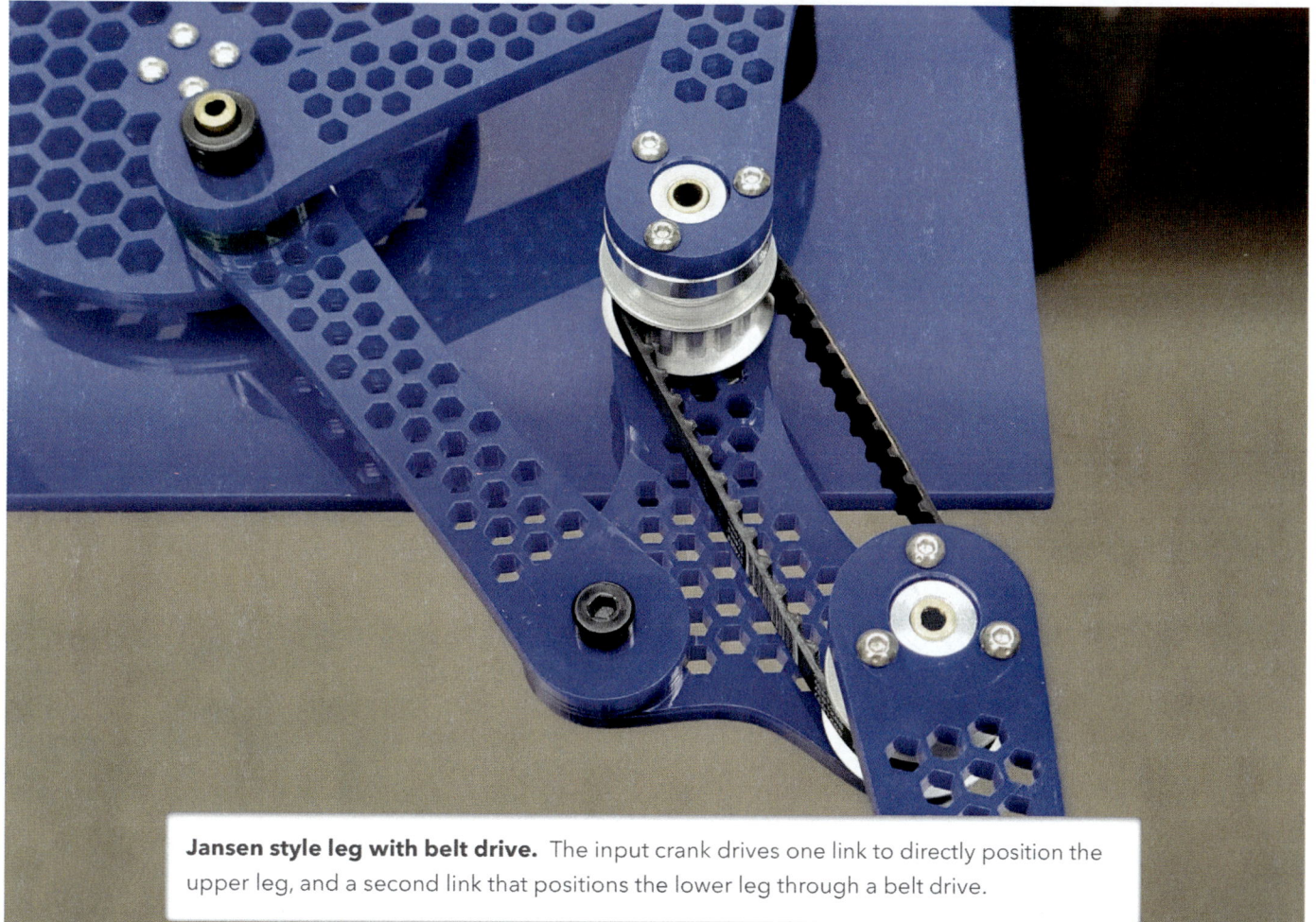

Jansen style leg with belt drive. The input crank drives one link to directly position the upper leg, and a second link that positions the lower leg through a belt drive.

Summary
Results of the Design Studies

The mechanical walkers described here are the result of design studies performed by teams of undergraduate and graduate students exploring the impact of a variety of design features on walking performance of the system

The six legs of the Anteater walker provide a nice tripod walking gate. With the equivalent of 10 joints for each of the legs, nylon bushings and PTFE washers were needed to reduce friction. And, the seven 3D printed gears needed lubrication; but the result was a smooth and stable walking movement.

The Nightmare and Chameleon walkers were designed and built in parallel after the Anteater. The focus of the nightmare walker was on the use of nylon shoulder bolts for the joints. The fewer number of legs, simpler leg mechanism with seven joints per leg, and nylon on acrylic joints reduced the friction in the system. This walker worked very well.

The Chameleon was constructed from acrylic with brass tube joints, and like the Nightmare walker benefited from less joint friction. However, the two motor system, with the two gear drives, universal drive shaft and vertical hinge for steering yields a heavy walker. The result was a nice walking movement with an effective steering capability. walkers worked well.

Steering electronics was the primary design focus for the two versions of the two legged Ostrich walkers. These designs demonstrated effective steering under RF control, however they also showed the importance of placement of the center of gravity to achieve effective walking in these systems.

Project-Based Learning

The Delorean walker introduces a new leg design using a four-bar linkage that drives a pantograph to scale up and position the foot trajectory. Wood was used for the chassis and leg mechanism to increase stiffness. Nylon bushings and acrylic spaces were effective in reducing friction. This walker used the electronics of the Ostrich walkers to control the speed of the drive motor controlled using an infrared transmitter and receiver.

The design of mechanical walkers provides a nice framework for a project course in mechanical systems, in which students learn the kinematic synthesis of linkages, digital prototyping and fabrication. Geogebra provides a convenient tool for the geometric constructions that yield effective leg mechanisms. Detailed digital modeling allows motion simulation as well as insight to fabrication. Finally, the parts are laser cut, 3D printed and assembled from readily available components. The electrical systems are easy to assemble and

Mechanical Flyers and Swimmers

program. Integration and testing results in walkers that move in a satisfying way along the ground.

The linkages used for leg mechanisms can also be used as drive mechanisms to flapping winged flying robots, as well as for articulated arm and leg movements for swimming robots.

Geogebra models and digital prototypes show that it possible to adjust a Jansen style leg mechanism to achieve an effective flapping

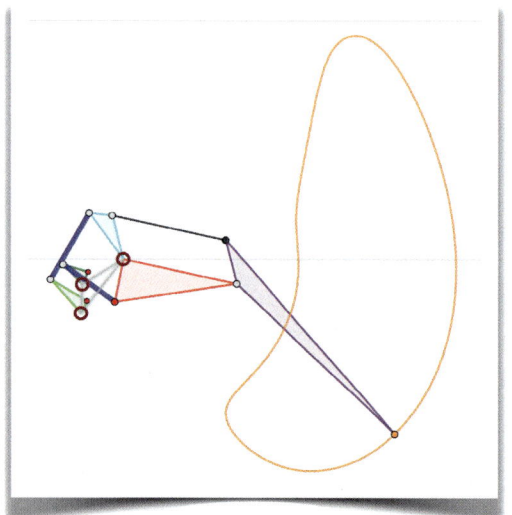

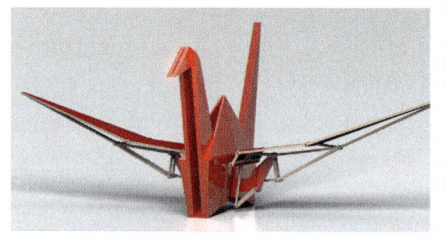

Digital model of wings for an origami Crane.

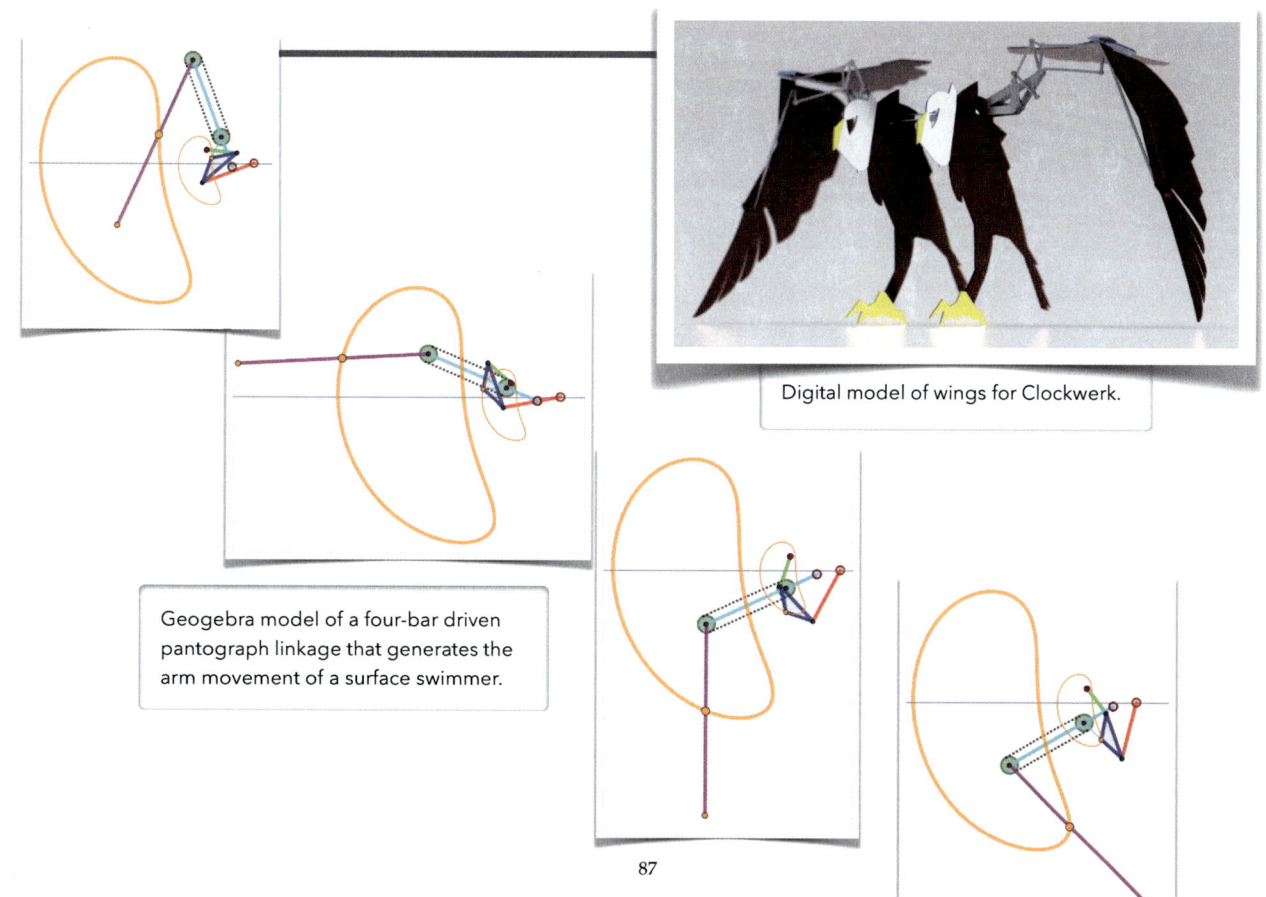

Digital model of wings for Clockwerk.

Geogebra model of a four-bar driven pantograph linkage that generates the arm movement of a surface swimmer.

Final Comments

wing drive mechanism for a large bird. Similarly, the geogebra model of the four-bar driven pantograph yields an effective articulated arm movement for a robot swimmer. We are exploring the application of all of our leg mechanisms for use in flying robots and swimming robots.

The mechanical walking robots described here range in size from 45 to 75 cm in length and 25 to 30 cm in width and they move at 20 to 30 cm/sec. The number of joints in the walker defines its complexity. The single drive motor keeps the overall system simple, even with a gear train that connects the motor to the input cranks for the legs. The electronic components and software required for speed control is also relatively simple, and does not increase significantly in complexity with the addition of a steering motor.

The design and fabrication of these walkers require solid modeling software to design the system, a laser cutter and some 3D printing to manufacture parts, hardware that is available on-line and to only hand tools for assembly. These devices provide a balanced engineering and manufacturing experience that develops mechanical, electrical and programming skills and results in a satisfying mechanical walking robot.

A project course centered on the design of these walkers can expect a novice team of four students to achieve the design and fabrication of a two legged walker within six to eight weeks. A skilled team of students can design and build a two legged walker and then a four legged walker in 10 weeks.

These walkers have sizable payload space that can be used for sensors, deployable structures, containers and communications equipment that extend and specialize their capabilities. Our hope is that this book provides the basis for many enjoyable experiences in the design of mechanical walking robots.

References

1. J. E. Shigley, 1960, The Mechanics of Walking Vehicles, Report No. RR LL-71, Land Locomotion Laboratory Research and Engineering Directorate, U. S. Army Ordnance Tank-Automotive Command 1501 Beard, Detroit Michigan, September.

2. J. M. McCarthy, 2019, Kinematic Synthesis of Mechanism—A Project Based Approach, MDA Press.

3. E. A. Dijksman, 1976, Motion Geometry of Mechanisms, Cambridge University Press.

4. K. Chen and J. M. McCarthy, 2021, Kinematic Synthesis of a Modified Jansen Leg Mechanism. In: Lenarčič J., Siciliano B. (eds) Advances in Robot Kinematics 2020. ARK 2020. Springer Proceedings in Advanced Robotics, vol 15. Springer, Cham. doi: 978-3-030-50975-0_30

5. T. Jansen, 2016, The Great Pretender, 3rd Ed, Nai010 Rotterdam.

6. W. B. Shieh, 1996, Design and Optimization of Planar Leg Mechanisms Featuring Symmetrical Foot-Point Paths, PhD Dissertation, The University of Maryland.

Index

Arduino Uno, 31, 76
Arduino Nano, 15, 60

Baltic birch, 7, 17, 69, 78
Battery, 13, 31, 49, 60, 69
Bevel gear, 55, 65
Brass tube, 10, 34, 40, 49, 76, 83

Chassis, acrylic, 22, 40, 52
Chassis, wood, 7, 69
Chen, K., 4, 89
Cytron shield, 76

Dijksman, E. A., 3, 89

Feet, 19, 37, 52, 66, 81
Flyers, 86
Foot trajectory, 3, 4, 24, 41, 54, 71
Four-bar linkage, 3, 7, 9, 41, 54, 84
Four-bar coupler curve, 3, 9, 39, 41, 51, 54
Four-bar driven pantograph, 9, 87

Gears, acrylic, 28, 57, 62
Gears, polycarbonate, 44, 57, 63, 74, 79

H-bridge, 15, 18, 31, 60
Hinge, vertical, 22, 52, 57, 84

Hub, 11, 18, 28, 34, 42, 49, 65, 73, 80

Infrared (IR) transmitter, 4, 76
Infrared (IR) receiver, 4, 76

Laser cut acrylic, 23, 28, 33, 46, 48, 62
Laser cut Baltic birch, 6, 17, 68, 78
Leg mechanism, Jansen style, 24, 71, 83, 86
Leg mechanism, pantograph, 3, 7, 9, 87
Leg mechanism, rectilinear, 3, 52, 54, 76, 66
Leg mechanism, skew pantograph, 3, 41

Jansen, T., 2. 4, 24, 68, 71, 89

McCarthy, J. M., 3, 4, 24, 89
Motor, drive, 4, 28, 31, 57, 55, 85
Motor, steering, 3, 52, 57, 60, 88

Parts list, 18, 34, 35, 49, 65, 80
Project-based learning, 85
PTFE (polytetrafluoroethylene) 73, 84

Radio-frequency (RF) transmitter, 28, 60
Radio-frequency (RF) receiver, 4, 18, 31, 60

Servomotor, 3, 22, 28,

Shigley, J. E., 3, 89
Shaft collar, 10, 17, 34, 49, 63, 73, 80
Shieh, W. B., 3
Shoulder bolt, 10, 42, 49, 84
Six-bar linkage, 41, 54, 62
Spacers, ABS, 40, 48, 79
Spacers, acrylic, 17, 47, 62, 73
Spacers, PLA, 64
Swimmer, 86, 87, 88

Timing belt, 11, 26, 35,

Universal drive shaft, 55, 56, 65

Walker, Anteater, 69
Walker, Chameleon, 52
Walker, Delorean, 7
Walker, Four legged, 38, 49
Walker, Nightmare, 38
Walker, Ostrich, 20
Walker, Two legged, 5, 20
Walker, Six legged, 67
Wing mechanism, Clockwerk, 87
Wing mechanism, Crane, 86

About the Authors

J. Michael McCarthy is a Distinguished Professor of Mechanical and Aerospace Engineering at the University of California, Irvine. He has published many papers and several books on the design of innovative articulated mechanical systems.

While a Visiting Professor at Stanford University in 2018 he began applying the theory of kinematic synthesis to the design of mechanical walking robots. Since then he has guided over 200 students in the design and fabrication of innovative mechanical walkers. He is now applying this theory to the design of mechanical flyers and swimmers.

Kevin Chen received his Bachelors of Science degree from California State University, Long Beach. He obtained his Masters of Science degree at the University of California, Irvine working on the design of mechanical walkers. His paper on Jansen-style linkages showed that Theo Jansen's innovative leg mechanism can be generalized to a broad class of mechanisms driven by two four-bar function generators. He has continued to design, model and build innovative two, four and six-legged walkers. His current focus is on adding wireless electronic systems for speed control and steering of these walkers.

Descriptions and Credit

Front Cover. Two legged horse walker designed and built by Koal Brockman, Alex Munoz, Evan Woo and Mirella Cruz in the Spring 2021.

Chapter 1. Geogebra image of the inflection circle and cubic of stationary curvature used to design a four-bar linkage with a flat-sided coupler curve and the static and dynamic models of the Jansen style leg mechanism by J. M. McCarthy.

Chapter 2. The Delorean walker was designed, modeled and built Kevin Chen in Spring 2021.

Chapter 3. The two Ostrich and chariot walkers were designed, modeled and built by James Le, Jason Lai and Matthew Gelacio in Winter 2021.

Chapter 4. The Nightmare walker was designed, modeled and built by Angela Cardamone in Fall 2020.

Chapter 5. The Chameleon walker was designed, modeled and built by Myia Dickens, Justin Lin, German Diaz, Dylan Salcido and Jeremy Jiang in Fall 2020.

Chapter 6. The Anteater walker was designed, modeled and built by Kevin Chen in Spring 2020.

Chapter 7. The rendering of a belt driven Jansen style leg mechanism was a design study by Kevin Chen.

— The collection of walkers was designed and built in Spring 2021: Four legged snail by Braydon Snow, Zack Whitrock, Jasper Shen, Felix Aguayo; Four legged turtle by Evan Woo, Mirella Cruz, Alex Munoz, Gunner Brockman; Four-legged elephant by Shaw Chiou, Jason Lai, Robert Laviguer, Taylor Orzuna; Four legged dinosaur by Pengpeng Wang, Kimberly Martinez, Andrew Diaz, Cameron Lee; Four legged dog by Vikram Bhave, Ken Nguyen, Tristan Cortez, Robert Flynn; Two legged horse by Koal Brockman, Alex Munoz, Evan Woo and Mirella Cruz; and Two legged elephant by Shaw Chiou, Minsoo Choi, Matthew Ives, Jason Lai.

— J. M. McCarthy reworked the Jansen style leg mechanism to be a wing mechanism, which was used to design digital prototypes in Spring 2021: Clockwerk by Matthew Ives and Matthew Gelacio; and the Crane by Darren Leung and Alyssa Noma.

— J. M. McCarthy reworked the four-bar driven pantograph to be the arm movement of a swimmer.

Back Cover: Five four-legged walkers and two two-legged walkers pictured in Chapter 7.

Photographs and renderings are by Kevin Chen and other graphics by J. M. McCarthy.

Books by J. Michael McCarthy

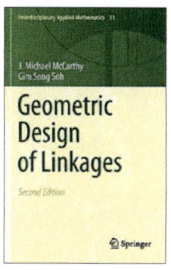

Geometric Design of Linkages, J. Michael McCarthy and Gim Song Soh. 2nd ed. 2011 edition, Part of: Interdisciplinary Applied Mathematics, Springer (December 2, 2010).

This book is an introduction to the mathematical theory of design for articulated mechanical systems known as linkages. The focus is on sizing mechanical constraints that guide the movement of a work piece, or end-effector, of the system. The function of the device is prescribed as a set of positions to be reachable by the end-effector; and the mechanical constraints are formed by joints that limit relative movement. The goal is to find all the devices that can achieve a specific task. Formulated in this way the design problem is purely geometric in character.

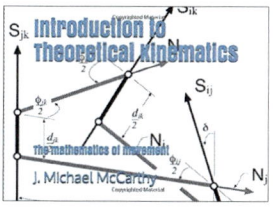

Introduction to Theoretical Kinematics: The mathematics of movement, J. Michael McCarthy. MDA Press (July 20, 2018).

An introduction to the mathematics used to model the articulated movement of mechanisms, machines, robots, and human and animal skeletons. Its concise and readable format emphasizes the similarity of the mathematics for planar, spherical and spatial movement. A modern approach introduces Lie groups and algebras and uses the theory of multivectors and Clifford algebras to clarify the construction of screws and dual quaternions.

Kinematic Synthesis of Mechanisms: A project based approach, J. Michael McCarthy. MDA Press (March 24, 2019).

This book is an introduction to the geometric theory used to design linkage systems that are critical components of machines ranging from vehicle suspensions to robot arms. The focus throughout is on graphical synthesis of linkages to control the movement of the legs for a walking machine. Increasingly complicated walking machines obtained from patent drawings, art and technology are used to motivate the theory.

In this book, we present the detailed design of mechanical walking robots that are driven by a single motor. These walkers rely on specially designed leg mechanisms coordinated by gear trains in order to walk, rather than multiple computer controlled motors per leg. The result is a simplified walking robot that provides a platform for other mechanical and electronic functions.

Two, four and six legged walkers are presented that implement different types of leg mechanisms and power trains. In each case, we provide drawings for a laser cut wood or acrylic chassis, 3D printed parts and a complete parts list. Several of the designs implement electronic components and software for speed control as well as an additional motor for steering.

Our goal is to provide enthusiasts of all backgrounds what they need to build a walking robot at home, to explore new design ideas, and, perhaps, to enjoy the operation of one of these robots as it moves across the ground.

The walkers we describe are the result of design studies by teams of undergraduate and graduate students in the Department of Mechanical and Aerospace Engineering at the University of California, Irvine. This book is dedicated to their creativity, commitment, and enthusiasm for the design of mechanical walking robots.

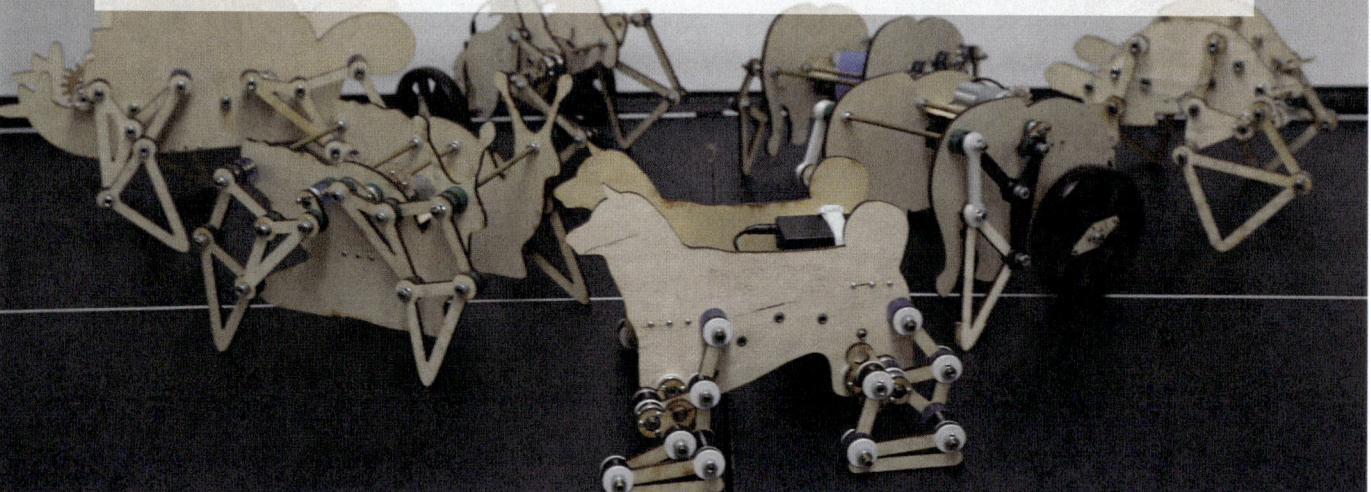

Made in the USA
Las Vegas, NV
18 March 2025